OXFORD
UNIVERSITY PRESS

Oxford University Press is a department of the University of Oxford. It furthers
the University's objective of excellence in research, scholarship, and education
by publishing worldwide. Oxford is a registered trade mark of Oxford University
Press in the UK and certain other countries.

Published in the United States of America by Oxford University Press
198 Madison Avenue, New York, NY 10016, United States of America.

© Oxford University Press 2019

CIP data is on file at the Library of Congress
ISBN 978-0-19-090741-9

1 3 5 7 9 8 6 4 2

Printed by Sheridan Books, Inc., United States of America

Chance in the World

A Humean Guide to Objective Chance

CARL HOEFER

ICREA and University of Barcelona

OXFORD STUDIES IN PHILOSOPHY OF SCIENCE

General Editor:
Paul Humphreys, University of Virginia

Advisory Board
Anouk Barberousse (European Editor)
Robert W. Batterman
Jeremy Butterfield
Peter Galison
Philip Kitcher
Margaret Morrison
James Woodward

Chance in the World

Contents

Boxes

Analytic Table of Contents

Chapter 1: Metaphysical Preliminaries

The chapter starts with a rough first approximation to the theory of chance to be developed. Objective chance facts are grounded on the existence of *patterns* in the events found in our world's Humean Mosaic; and the chance facts so grounded will later be seen to be apt for guiding rational credences (subjective probabilities) in the way captured by the Principal Principle (PP), which in one form can be written:

$$PP: Cr(A \mid XE) = x,$$

where X is the proposition that the objective chance of A is x, and E is background knowledge.

After laying out the basic idea, the rest of the chapter is dedicated to exploring two issues. First, what is this Humean Mosaic (HM)? What does it contain, and what does it leave out? What understanding of time is presupposed? What sorts of properties and things make up the events in the HM? Unlike Lewis and other Humeans, I do not insist that the HM have only "occurrent properties" in it; I am happy to allow ordinary dispositions, causal relations and/or causal powers, and suchlike things—as long as they are not *probabilistic*. I also adopt an agnostic or ecumenical attitude concerning the laws of nature: they may be Humean facts supervening on the HM, or primitives, or even necessary truths of some sort—again, as long as they are not probabilistic.

Second, the rest of the chapter gives an extended discussion of the idea of considering objective chance facts to be *primitives* of some sort, as most propensity views hold, or to be based on primitively (irreducibly) chancy laws of nature. I develop what I call "the dialectics of primitive chance," an extended attempt to explore *what it could mean* to postulate primitive chances or chancy laws. In the course of the dialectics we see that it is by no means clear that any answer can be given to this question of the meaning of primitive chance claims. But we see also that a tacit—and illegitimate—invocation of the PP helps explain why philosophers often *think* they understand the meaning of primitive chance claims. The invocation of the PP is illegitimate, though, because there is no way to show that a bare primitive posit deserves to guide credence in the way captured by PP. The dialectics lead, then, to the conclusion that we don't really understand what it could mean to postulate primitive chances or chancy laws. Thus, the chapter prepares the reader to be receptive to an alternative approach to understanding objective chance, one that is explicitly reductive and that eschews reliance on mysterious metaphysical notions.

Chapter 2: From Lewisian Chance to Humean Chance

In this chapter I introduce David Lewis' highly influential and much-discussed views on objective chance. First I look at his 1980 treatment of objective chance and the Principal Principle, and criticize some of the decisions Lewis made in that paper, especially concerning time and the notion of "admissibility." I then go on to present his final, 1994 theory of objective chance. Lewis adapted his Best Systems Analysis (BSA) of laws of nature in order to allow it to accommodate probabilistic laws. His final proposal, then, is that the objective chance facts in our world are whatever chance facts are to be found in the Best System of laws for our HM. This has an unfortunate consequence: if the Best System of laws for our world has no probabilistic laws in it (e.g., because some deterministic physical laws are able to cover all events in the HM perfectly well, and the Best System contains only those laws), then there are no objective chances in our world. But we clearly do believe there are at least some objective chances: of coin flips and dice rolls, of radioactive elements decaying in a certain time, of $spin_z$-measurements yielding the result "up", etc.

I argue that Lewis was mistaken to maintain that non-trivial objective chances could not coexist with determinism at the fundamental-physics level, and offer a detailed diagnosis of where he went wrong in his treatment of chance. Lewis erred in making objective chance a time-indexed notion and in saying that "what is past is no longer chancy"; he erred in assuming (sometimes) that every proposition should have some objective chance of being true (i.e., that the domain of the chancy is all facts); and he erred in linking his account of chance too tightly to his BSA account of laws. But two key pillars of Lewis' approach remain valid. First, his contention that the ability to *demonstrably* play the chance role captured by the PP is crucial for any account of the nature of objective chance. And second, his intuition that a Best System approach in which chance facts supervene on patterns in the HM is a promising approach vis à vis allowing such a demonstrable grounding of the PP. The next chapters will develop these pillars into a new Humean theory of objective chance.

Chapter 3: Humean Objective Chance

In this chapter the core of my theory of Humean objective chance (HOC) is laid out and discussed using a number of examples. Being a Humean account, in which the chance facts supervene on facts about the patterns of events in the actual HM of our world, the account can be thought of as a sophistication of actual frequentism. But the differences are substantial! In particular, the Best System aspect of the theory lets it overcome two serious problems (and several less serious ones) of simple actual frequentism: the reference class problem, and the problem of few/no instances.

The theory can be summarized as follows: Chances are constituted by the existence of patterns in the mosaic of events in the world. These patterns are such as to make the adoption of credences identical to the chances rational in the absence of better information, if one is obliged to make guesses or bets concerning the outcomes of chance setups (as will be proven in chapter 4). Some stable, macroscopic chances that supervene on the overall pattern are explicable as regularities guaranteed by the structure of the assumed chance setup, together with our world's

Chapter 4: Deducing the Principal Principle

rational credence" (1994, p. 484). In reply, Michael Strevens and Ned Hall have claimed that Lewis was mistaken, as either there is *no way at all* to justify PP, on any view of chance (Strevens, 1999), or no way for a Humean to do the job (Hall, 2004). This chapter proves that Lewis was right, by giving two distinct justifications of the PP for HOCs.

The first justification can be thought of as "consequentialist" in nature: it tries to show that in practical decision-making, an agent who has to make repeated bets on whether a certain type of chance setup will yield result A, and who knows the chance of A but has no inadmissible information (the scenario of PP), will do better setting her credence equal to the chance of A than she can do with any other, significantly different, betting strategy. The consequentialist argument, while useful and able to overcome Strevens' objections to such arguments, has certain limitations. Fortunately, the second argument does not suffer the same limitations.

The second way of justifying the PP for HOCs is an *a priori* argument, rather than a consequentialist one. That is, what it demonstrates is that an epistemic agent meeting the conditions for application of PP is irrational—*logically* incoherent, in fact—if she sets her credence to a level substantially different from the chance. This argument is an adaptation of one originally offered by Colin Howson and Peter Urbach (1993) to justify the PP for von Mises–style hypothetical frequentism. In chapter 1 I showed that the argument is actually problematic for hypothetical frequentism, because of the infinities involved in the latter, but no such difficulty arises when the argument is used in the case of HOCs. The discussion of the two justifications of PP is rounded out by considering whether certain loopholes or minor *lacunae* exist in the arguments; I argue that there are no *significant* loopholes in the consequentialist argument, and insofar as there are minor ones, the a priori argument serves to fill the gap.

In section 4.2, I briefly discuss the epistemology of HOC, arguing that there is no *special* difficulty in coming to know what chances are to be found in the Best System for our HM. Ordinary inductive reasoning, of the type we inevitably rely on in all science and in daily life, is enough to make the epistemology of HOC transparent. As long as one is willing to set aside Hume's skepticism about induction in general—as we all are willing to do, in all contexts *except* philosophical discussion of that very issue—there is nothing especially difficult in the epistemology of Humean chance. In section 4.3 I briefly survey how other accounts of objective chance fare with respect to justifying PP. Finite frequentism fares even better than HOC in the *a priori* argument; but its epistemology is much more problematic and its other defects render this small advantage moot. Hypothetical frequentism fares badly, as noted earlier, because of infinity problems; and if one tries to modify or fix the view so that it can justify the PP, it becomes very nearly identical to HOC. Primitive or propensity-type accounts fare worst of all, as there is no non-question-begging way for them to even attempt to show that PP is rational for primitive chances (as we already noted in chapter 1). Finally, Sober's No-Theory theory of objective probability does somewhat better than propensity accounts, but only in cases where the probabilities are specified directly by a scientific theory (such as quantum theory, or statistical mechanics).

Chapter 5: Undermining

A large proportion of the discussion of Lewis' approach to objective chance has centered around the problem of *undermining* and the apparent contradictions to which it may give rise, when Humean chances and the Principal Principle come together. HOC is subject to the undermining problem just as much as Lewis' account was. So the aim of this chapter is to give a full discussion of, and resolution of, the undermining/contradiction problem for Humean chance. First, I lay out the problem and discuss the attempts by some authors (including my earlier self, 1997) to dismiss the problem as illusory or based on a mistake in reasoning. Unfortunately, all these attempts are flawed, and the problem is real.

Using the insights gained in this review of the literature, I show that the correct way to overcome the undermining problem is *via* a revised form of the Lewis-Hall response, which involves making a small amendment to the PP itself. Lewis and Hall argued that the correct rational principle, NP ("New Principle"), is much like PP but involves conditionalizing on the truth of the full chance-theory T_w (in my terms, the truth of the Best System of chances for our world):

$$\text{NP: } C(A \mid T_w E) = x = \Pr(A \mid T_w)$$

I argue that while this does solve the problem, it is overkill; all that a reasonable agent needs to conditionalize on is the non-occurrence of an undermining outcome in the chance setup:

Let $\{A_i\}$ be the set of possible outcomes we are contemplating, among which the subset $\{A_u\}$ are outcomes that, in conjunction with E, are underminers of the chance rule X. Then for all A_i,

$$\text{NP*: } C(A_i \mid XE) = \Pr_X(A_i \mid \neg\{A_u\})$$

This modified version of the PP is identical, in all *practical* scenarios, to the original PP, because in real life we will never be faced with a situation of needing to assign probabilities to events that could undermine the Best System's chances. But it is good nevertheless to have the undermining problem resolved, and I argue that adopting NP* as the official, always-applicable chance-credence principle does the job.

Chapter 6: Macro-Level and Micro-Level Chances

In this chapter, I tackle conceptual problems that may arise if we think of the Best System as giving us two (potentially) *different* chances for the same macroscopic event: one that arises by mathematical entailment out of chancy microphysics (something like quantum theory's chance rules), and a second one that is in the system because it supervenes directly on the pattern of events at the macro-level. Given the pragmatic approach of HOC, such dual-chance-value situations could exist, though there are formidable conceptual difficulties about how to derive probabilities for macroscopic event-*types* from micro-level chance laws. If such dual-value situations exist in our Best System, I argue, we would have reason to apply PP to the macro-level chance rules, but *not* to the micro-derived chances. The discussion here reveals a heretofore hidden limitation of both HOCs themselves, and the validity of the arguments from chapter 4 justifying PP. The limitation is one of no practical epistemic importance, since finite beings can never actually calculate the micro-derivable objective chances under discussion here.

Chapter 7: Humean Chance in Physics (co-authored with Roman Frigg)

As we have seen in the preceding, I adopt a pragmatic and ecumenical approach to objective chance, i.e., one that is open to the Best System of chance rules containing rules phrased in terms of the kinds and events of everyday macroscopic life, as well as chance rules of the kind we might expect fundamental physics to provide. Nevertheless, some of the most compelling examples of the existence of truly *objective* probabilities come from physics, in particular quantum physics and statistical mechanics. So it is crucial to the overall success of HOC that it be compatible with the objective probabilities we find in modern physics.

Although probabilities are nowadays rather ubiquitous in physics, they basically can be classified into two distinct kinds: purely stochastic probabilities (not grounded on anything else), or probabilities that are superimposed on a deterministic underlying theoretical structure. We will discuss how HOC fares as an account of two paradigmatic types of objective probability, one of each kind. First we will discuss, in some depth, objective probabilities in classical (Boltzmannian) statistical mechanics (SM). We will show that HOC does capture the central probabilistic postulates of SM, and indeed that it may do so in two distinct ways. Along the way, we make some critical remarks concerning other approaches to capturing SM probabilities as objective chances. Second, we discuss standard, non-relativistic quantum mechanics (QM), where the notion that fundamental physics is at bottom chancy first became widely accepted. We explain why HOC is especially apt for capturing the probabilities of QM; other accounts *may* do equally well (though some clearly do not), but none can do the job better.

Chapter 8: Chance and Causation

Causality and objective probability are often linked, and the links may go in either direction. For example, some philosophers have tried to characterize objectively chancy setups as incomplete, *partial* causes of the various possible outcomes the setup may yield. Other philosophers have proposed probabilistic theories of causation, defining a cause *c* for an effect *e* as a factor whose presence raises the objective probability of *e*.

I do not subscribe to any reductive or quasi-reductive link between causation and objective chance. Nonetheless, it is clear that there is *some* link between causation and probability, as can be seen in the following example which I use as a foil for this chapter. Lisa and Bob both work at a downtown advertising firm, and Bob just gave a presentation of an ad campaign to some important clients. Lisa says: "Great presentation, Bob! You really upped your chances of getting that promotion."

Taken at face value, Lisa seems to be asserting that there was some earlier objective chance that Bob would soon get his promotion, and that after his presentation that chance now has a higher value. But for various reasons, this is a bad way of capturing what Lisa is saying. Discussing this example in some depth leads us to see that the right way of rephrasing Lisa's remark is in terms of *subjective* probability: her credence in the truth of "Bob will get his promotion" has gone up because of his great presentation. Generalizing from this case, I propose that the strongest general principle that links causation and probability is a Cause-Probability Principle (CPP), which says (roughly) that when an agent learns that a cause *c* for an effect *e* has been introduced or put into action, then her subjective probability for the occurrence of *e* (if she has one!) should be at least as high as it was beforehand (that is, should go up in general, and in any event not go *down*). This principle, which is by no means very important or illuminating, is nonetheless (I claim) the strongest *general* link that we can make between causation and probability. As with the PP, a more careful statement, including an "admissibility" clause, is needed in order to have a defensible CPP. I end the chapter with some general remarks about causation that I take to be plausible in light of the preceding discussion.

Preface

"Chance" is a word we use in a variety of ways and contexts. Often we speak of something happening "by chance," meaning that it was unplanned, or a coincidence, or "for no reason" compared to other similar events that have tidier explanations. These uses of "chance" do not presuppose the applicability of probability theory to the event in question in any way, and they will not be our concern in this book. A second sort of use of "chance" has become commonplace, exemplified by statements like "If you smoke, you increase your chance of getting lung cancer."

Here it is not clear whether numerical probabilities are lurking in the background, or not. Might we be able to correctly say something like this, without committing ourselves to there being a certain chance (numerical, objective probability) of your getting lung cancer if you don't smoke, and another, higher one if you do? This is an important question that will be addressed in chapter 7.

Finally, we sometimes also say things like "The chance of getting a 6 on rolling a fair die is 1/6." Uses of "chance" like this clearly do invoke the mathematical notion of probability, and most of us believe that such statements are (often, at least) meant to be assertions of the existence of certain *objective probabilities*—that is, probability facts made true by what is "out there" in the world, not just in our heads. It is this sort of use of "chance" that is our main topic in this book.

I will follow common recent practice by using interchangeably the terms "objective probability" and "objective chance" or simply "chance." But as we will see in chapters 1 and 3, this practice is potentially controversial. There are a number of philosophers who wish to reserve the label "chance" for something more modally or ontologically robust than some other types of probabilities that they are willing to countenance and consider "objective." For those philosophers, the account of "objective chance" to be developed here might better be called a theory of "objective probabilities," "general probabilities," or even "objective epistemic probabilities." But I stand by my terminology, for reasons that will emerge gradually in the sequel. A sufficient defense of my terminological practice can be found already in chapter 1, though the more complete defense will be the success of the arguments of the book as a whole, particularly chapters 1 through 4.

The goal of this book is to sketch and defend a new interpretation or "theory" of objective chance, one that lets us be sure such chances exist and shows how they can play the roles we traditionally grant them. It is a work of metaphysics and ontology: I will argue that chances are indeed to be found *in the world*—the actual world, and not merely in counterfactual infinite-trial worlds, or in models, or in our minds. But it is also a work of philosophy of science. By uncovering exactly what chances are, in the world, we will gain an improved understanding of how and why objective probabilities are used successfully in many sciences; and what pitfalls may be lurking in our practices, if they are founded on incorrect views about chance.

Why is a new theory needed? Prior to the work of David Lewis on objective chance, especially his definitive theory presented in (1994), the philosophical discussion of objective probability had hardly changed in half a century. If one compares Wesley Salmon's discussion of interpretations of probability in his classic textbook (Salmon, 1967) *The Foundations of Scientific Inference* with Alan Hájek's (2008) entry in the *Stanford Encyclopedia of Philosophy*, one will be more struck by the similarities and continuities, than by any major differences. Objective probabilities might be understood, we are told, as primitive causal *propensities*; or as *the limiting frequency of outcomes in a hypothetical infinite sequence of trials*; or as simply the *actual frequency of outcome-type divided by number of trials, using appropriately chosen reference classes*; or (mentioned mainly out of respect for Carnap and Keynes) as *degrees of partial entailment*. There have been a number of "hybrid" accounts proposed in the past few decades, especially views attempting to "objectify" subjective probabilities in one way or another, but none has earned widespread acceptance or freed itself from the key defects of the classical accounts. In the years since 1994 a number of new theories of objective chance or objective probability have been put forward, but many have been limited to certain specific domains (e.g., quantum mechanics or statistical mechanics), and none has garnered widespread acceptance.

The stagnation of philosophical discussion of objective chance for decades would be understandable, if at least one of the standard views were tenable and unproblematic; but in fact, in the judgment of most authors writing such overviews, this is not at all the case. *None of the standard views is free of grave epistemic and/or metaphysical problems, other than the actual frequency account; and that account is held to be completely ruled out* by reason of the reference class problem and clashes with basic intuitions.

Perhaps this state of affairs explains the large degree of interest stirred by Lewis' presentation of a truly new analysis of objective chance. My own interest in probability certainly stems from that work, which I had the pleasure to see Lewis present at the University of California, Irvine, in 1993. After corresponding briefly with Lewis concerning that 1994 paper, I grew more and more interested in his approach to objective chance, which seemed to offer a way around most of the difficulties of the traditional accounts.

The subtitle of this book emulates the title of Lewis' seminal 1980 paper "A Subjectivist's Guide to Objective Chance"—while indicating an important difference in perspective. The view I will defend shares two major tenets with Lewis' account of objective chance:

(1) The Principal Principle (PP)—a principle that links rational degrees of belief to objective probabilities—tells us most of what we know about objective chance;

(2) Objective chances are not primitive modal facts, propensities, or powers, but rather facts entailed by the overall pattern of events and processes in the *actual* world. Thus, the account to be developed here is compatible with Humean metaphysical strictures (though, as I will show, it does not require them.)

But as we will see in chapters 3 and 4, the account I favor disagrees with Lewis' on a host of other points.

Another subtitle I considered was "A Skeptic's Guide to Objective Chance." But while the account of chance to be developed rejects naive or brute realism about objective chance—i.e., a primitive/ dispositional story about chance—it is certainly not anti-realist overall. The world contains contain objective chances, just as it contains books and galaxies and cities, even though these things are (arguably) not *fundamental* entities or properties. My rejection of chance-primitivism has some consequences, as we will see, for what laws of nature cannot be and what causality cannot be; but nothing beyond those consequences is presupposed. In particular, I do not presuppose (or endorse) Humeanism about laws of nature or causality. Nevertheless, once we begin to discuss the positive account on offer, it will be referred to as "Humean objective chance," or HOC for short, because it *is* a perfectly Humean account of chance.

By rejecting more metaphysically loaded approaches to chance, my approach should have appeal to subjectivists and other skeptics who might, deep down, like to have a story to tell about objective probabilities, but find themselves repelled by the existing options. Conversely, by offering a genuine *analysis* of objective chance, my approach will have appeal for proponents of robust theories of objective probability who are unable to rest easy with pure subjectivism or with any of the traditional accounts.

By the end of chapter 3, when my account of objective chance is fully laid out, I suspect that many readers will think to themselves something like this: "Well, this so-called new theory is really extremely close to what I've thought all along; it's just a variant of a __ _____ account." In the blank, those readers will insert their favored approach's name: propensity, long-run frequency, actual frequency, epistemic, objective Bayesian, no-theory theory, and so on. If this occurs, then I am on the right track. For how could *all* philosophers have been simply mistaken or confused about the nature of objective probability, in the past? They couldn't have been, and have not been. In one way or another, by my lights, most extant accounts of the nature of chance get *something* right; and sometimes quite a lot right. As I see it, the best account of chance should be one that makes clear how and why earlier thinkers got a lot right in their accounts of chance. Over the course of the first four chapters we will see how my Humean account meets this goal.

The key to understanding what chance, as we will see, lies in understanding what objective chance is *good for*, i.e., why we have the concept and what work it does for us. And what chance is good for is precisely what PP captures. PP says, in words, something like this: An ideally rational agent who knows that the objective chance of A is x, and has no other information bearing on whether or not A (so-called *inadmissible* information), will have *credence* or *subjective probability* for A equal to x.

We will discuss in depth the meaning of all the terms introduced in the preceding, in the course of the book; but for now, the gist of PP should be clear enough. A rational agent who takes herself to know the objective chance of A and nothing further that bears on A's truth or falsity will use the objective chance x for her personal betting odds, make decisions about actions using x as the relevant probability for A, and so forth. In terms of the coarser distinctions of folk psychology: if x is very low, then the agent does not expect A to be true, while if x is very high she does expect it (and in some circumstances we would naturally say the agent *believes* A); if x is near $\frac{1}{2}$, the agent is perfectly uncertain as to whether or not A; and so forth.

PP is presented as a principle of rationality, something that captures part of what it is to be rational or reasonable. To the extent then that one takes the principles of rationality to have a normative character, PP is normative. It takes "is"-type facts, facts about how things objectively are in the world, and connects them with facts about how agents "ought" to arrange their beliefs and actions. How can this is/ought connection be made? What sort of understanding of objective chance can make this connection justifiable, and ratify the intuitive correctness of PP? As we will see in the course of chapters 1–4, a Humean account of chance—and perhaps *only* a Humean account—is able to do these things, demonstrably and clearly.

As I mentioned earlier, I will be advocating a Humean approach to chance, but not (necessarily) to other related notions such as causation, laws of nature, psychology, or metaphysics in general. Humeanism about chance, in my sense, means *reductionism* about chance, of a certain sort. Facts about objective chance can be understood as reducing to certain sorts of facts about *what there is* and *what happens* in the world, understood in the most ontologically innocent way. There are, we think, tables and chairs, people and their brains; chromosomes, x-rays, and neutrinos. And there are events such as: 1,000 neutrinos passing through a single cell in a span of one second; 10 consecutive coin flips all landing Tails; and inflation in Germany exceeding 3% in 2020. Such things and events exist and are spread around in space and time. All the things there are, at all times, and the events that occur— all this makes up what I will call the Humean Mosaic (HM). The HM may be all there is, though I will not assume this. What might be left out of HM? For some philosophers, much that is important: things like dispositions and potentialities; pure universals and abstract objects; laws of nature; causings and necessitations (law-based or otherwise). These things may also be part of reality; I myself believe in some of them, and not others. But they are not needed for the correct analysis of objective chance. Chances can be reduced to facts about the HM; in particular, to facts about the patterns that exist and can be discerned in HM's events. As we will see, this is one great virtue of HOC. Since all philosophers will agree that the reduction base—the HM—exists, all must agree that Humean chances, *qua* complex facts about the world, exist (at least, if the recipe for deriving Humean chances from the HM is sufficiently well-defined). When in addition it is shown (chapter 4) that Humean chances permit the deduction of PP as a requirement of rationality, the case for HOC becomes compelling indeed.

◆

Acknowledgments

During the inordinately long gestation period of this book I have been supported and encouraged by countless friends, colleagues, and funding agencies. I owe special debts to the following people who encouraged me, read sketchy drafts, listened to numerous talks with a familiar punchline, disputed my arguments, told me when I was flat wrong, and generously conceded when I was right: Marshall Abrams, Craig Callender, Nancy Cartwright, Jose Díez, Nina Emery, Aldo Filomeno, Mathias Frisch, Alan Gibbard, Alan Hájek, Toby Handfield, Nick Huggett, Jenann Ismael, Jim Joyce, Marc Lange, Dustin Locke, Barry Loewer, Aidan Lyon, Genoveva Martí, Manolo Martínez, Tim Maudlin, Chris Meacham, Alex Meehan, Wayne Myrvold, Manuel Pérez Otero, John Roberts, Alex Rosenberg, Simon Saunders, Elliott Sober, Albert Solé, Mauricio Suárez, and Alastair Wilson. My debt to Roman Frigg is especially great, as he has helped me work out many aspects of my account of chance over the years, and especially its application to physical theories; we have co-authored three papers together, and he generously co-authored chapter 6 of this book as well. I also want to thank Christopher Evans and Jeimy Aquino for doing an excellent job of making the figures in the book clear and good-looking, and for equally excellent work preparing the index.

I owe a great debt of thanks to a number of close friends who, with patience and perseverance that can only be described as born out of love, never stopped encouraging me to finish the darn thing. But my most profound debt is to my family, who have blessed me with their love, support and encouragement throughout my career, and—above all others—to my wife and colleague Genoveva Martí. From the bottom of my heart I thank her for her wise advice, endless patience, tireless encouragement, and unconditional love, without which this book would never have come into being.

Chance in the World

1

Metaphysical Preliminaries

1.1. Humean Objective Chance (HOC):
A First Sketch

This chapter is mostly dedicated to metaphysical setup and ground-clearing, a labor that seems necessary in order to prevent later misunderstandings and, more importantly, to demonstrate the need for a reductionist account of objective chance. But before the setup and ground-clearing can be done, the reader needs to have a first-approximation idea of the theory to be developed and defended in later chapters. And most especially, the reader needs to understand the Principal Principle (PP) and its crucial role in our thinking about, and our uses of, objective chances.

The PP, in one simple form, can be given as follows:

Let $Cr(\_|\_)$ be a rational subjective probability function[1] (credence function), A be a proposition in the domain of the true objective chance function $P(\_\_)$, E be the rest of the agent's background knowledge, assumed to be "admissible" with respect to A, and X be the proposition stating that the objective probability $P(A)$ is x. Then:

$$Cr\left(A \mid XE\right) = x \qquad \text{(PP)}$$

---

[1] In (Lewis, 1980/1986a) the PP is actually specified as applying to a rational *initial* credence function, which can be thought of, as Alan Hájek likes to say, as a "Superbaby" credence function. It is a credence function over all propositions that obeys the probability calculus (and thus is, among other things, omniscient about matters of logic—hence "super"), and which has not yet been "updated" by conditionalization on experience.

We will discuss the meaning of all the terms introduced in the preceding in depth, in the course of the book, and credence (or subjective probability) is discussed shortly below, in Box 1.1. But for now, the gist of PP should be clear enough. A rational agent knows the objective chance of A and nothing further that bears on A's truth or falsity, will use the objective chance x for her personal betting odds, make decisions about actions using x as the relevant probability for A, and so forth. As (Lewis, 1980/1986a) stressed, PP captures much of the essence of our concept of *objective chance* because it captures what chances are *good for*, why we would like to know the chance-facts if we can. And like Lewis, I take it as the most important demand we should make of a *theory* of chance that it should make crystal clear why objective chances (as characterized in the theory) can and should "play the PP-role." That is, a good theory of objective chance has to make clear to us why the objective chance facts should constrain rational agents' subjective degrees of belief in the precise way expressed in PP. As we will see later in this chapter, this is a demand that at least two traditional theories of objective probability are unable to meet.

But, as Lewis thought he saw "dimly but well enough," a Humean reductionist theory of objective chance *is* able to meet the challenge of showing PP to be a genuine constraint on rational agents—*clearly* able, and perhaps *uniquely* able to do so (as we will see in chapter 4). So what is a Humean reductionist theory of chance? The core idea is this: the chance facts are reducible to certain facts about the *patterns* to be found in the whole panoply of events that happen in the history of our universe. What kinds of patterns? Broadly speaking, and unsurprisingly, they are stable and *stochastic-looking* or *random-looking* patterns. There is more to the story than just that, as we will see in chapters 2 and 3, but for now we can go on with this simple idea of what a Humean account of objective chance maintains: the fact that <the objective chance of A occurring in chance setup S equals x> is grounded on the existence of certain sorts of patterns in the whole panoply of events that happen in the universe's history.

Box 1.1 What Is Credence or Subjective Probability or Degree of Belief?

Throughout this book I will presuppose that readers are already familiar with the notion of an agent's *credences* or *subjective probabilities*. I will use these two phrases interchangeably, mostly using "credence" because it is shorter. It is typical, to convey this concept to those unfamiliar with probability talk, to say something like: "Your credence in a proposition *A* is your degree of belief that *A* is true; or how willing to bet on *A* you would be, i.e., what sort of payoff ratio you would want to have for a bet on *A*." But when these concepts are examined with a bit of care, it is not clear that either one is clear-cut and well understood (either the notion of degree of belief on its own, or the idea of a person's betting dispositions).

In a paper of great destructive brilliance,[a] Eriksson and Hájek (2007) equate credence with degree of belief, and they try to systematically dismantle existing attempts to give some sort of analysis of the notion in terms of betting quotients. Those attempts either define credence in terms of the odds at which an agent would be disposed to accept a bet on the truth of a proposition, or claim at least that betting behaviors serve to *measure* (perhaps imprecisely) credences. For example, de Finetti (1990, p. 75), proposed: "The probability P(E) that You attribute to an event E is therefore the certain gain *p* which You judge equivalent to a unit gain conditional on the occurrence of E. . . ."

Eriksson and Hájek point out a number of difficulties with trying to understand credences in terms of dispositions to make or accept bets. Some people are morally opposed to gambling, and so have no disposition to accept *any* bets on any proposition. Some enjoy risk-taking and are disposed to gamble at odds they know to be too high or low; some stoical or Buddha-like agents may be indifferent to all bets equally; and so forth. But these agents, intuitively, may perfectly well have definite degrees of belief in at least certain kinds of propositions, degrees that the standard betting analyses fail to capture.

These sorts of problems seem to me to be elegantly side-stepped by the sort of definition proposed by Howson and Urbach (1993), a definition that Eriksson and Hájek fail to consider. Howson and Urbach propose that for an agent's subjective credence in A to be p is for the agent to judge that there is no advantage to taking either side of a bet concerning whether A, at odds of $p:(1-p)$. In one fell swoop, the move of shifting the focus to what the agent considers *would be* fair betting odds (as opposed to *her own* actual (or counterfactual) betting dispositions) overcomes most, if not all, of the difficulties canvassed by Eriksson and Hájek. This definition of credence is put to work later in chapter 1, and in chapter 4 where I demonstrate the validity of the PP starting from my account of objective chance.

That said, it is not clear to me that even Howson and Urbach's definition captures fully the phenomenon of our subjective degrees of belief. My concern is that there are, perhaps, two largely overlapping but distinct senses in which one can have a strong or weak degree of belief in a proposition. One sense is the one captured by the betting analysis just discussed. The other has to do with the *strength of evidence that would be needed to convince us of the falsity (/truth) of the proposition at issue.* This sense comes to the fore for propositions such that it is more natural to speak of belief/disbelief *simpliciter* rather than degree of belief or credence. Everyday life is full of common examples: data CDs hold less than a gigabyte of data, you own exactly one car, solid gold is more dense than ice, etc. You probably have a very strong degree of belief in each of these propositions (at least, if you are a one-car person), and in fact we would normally express ourselves by saying "I believe that A" (or, in many contexts, "I know that A"), rather than with any talk about a *degree* of belief. Insofar as the notion of credence applies, it is surely no exaggeration to say that your credence is near to 1 in each of them. But not exactly 1. You recognize your own fallibility, and can conceive of becoming convinced that you were wrong about any or all of them. But it would take some extremely convincing evidence to do the trick, and if someone *did* try to convince you that one of these is false, you would suspect a practical joke was being played on you. Indeed, you might need the word of God herself to begin to accept it; then again, even that might not be enough, if you start from a low credence in the existence of such an entity. Similar examples can be given on the side of disbelief and extremely low credence.

The point is that there is a sense of "degree of belief" that comes into play for propositions we believe or disbelieve that has to do with how strongly we would tend to hold on to them in the face of (apparently) contradictory evidence, rather than with anything about what betting odds would be fair.[b] In fact, for such propositions one is apt to have nothing to say if asked for the *fair* betting odds on the proposition; there is no such thing as a fair bet on the truth of a proposition you *know* is false! If reminded of one's epistemic fallibility, one might say something like, "Oh, alright; I guess a bet at a billion to one odds would be 'fair.' But you won't get *me* to take the bet, bub!" High degree of belief, in this sense, does not seem to either correspond to, or be well measured by, hypothetical fair betting odds.

It is no great surprise that subjective probabilities, understood as something "in the head" of agents, may get mushy and not-well-defined when the strength of belief gets too close to either 0 or 1. I think they may also get mushy, or perhaps simply non-existent, when it comes to certain sorts of propositions about which we feel essentially clueless. To borrow an example from Kenny Easwaran: what is your credence in the proposition that the average American eats more than 10 whole pickles each year? When I introspect, I get nothing close to a number in my mind, even if I translate the question into one about fair betting odds. This illustrates a phenomenon that I suspect is important and universal (or nearly so) among actual human agents: we simply do not *have* credences, for many propositions that we can understand. (In fact, at times I am inclined to think that most instances where we feel we have reasonably sharp credences far from 0 or 1 are instances where we have invoked the PP (usually tacitly) from what we take to be some objective probability.) Such "credence gaps," if they exist, will cause no problem for anything we need to do with subjective probabilities in this book.

a With apologies to Anscombe (1971) and Russell (1912).
b Leitgeb (2017) defends an account of ordinary full-stop belief that equates it with robustly or stably high credence.

1.2. The Humean Mosaic

1.2.1. What Is the Humean Mosaic?

The Humean basis of this reductive story, the "whole panoply of events" just mentioned, is usually called the "Humean Mosaic," or HM for short. The HM can be thought of as, basically, the whole universe: everything that exists, everything that ever happens, past as well as future, in all parts of space. For David Lewis, the Humean Mosaic was something whose precise description we would expect physics (ideal, future physics) to provide. But Lewis had his own ideas about what it would be like in broad-brush terms, or in a simplified version compatible with classical (i.e., non-quantum) physics. The HM might consist of a 4-d continuum or "manifold" of spacetime points, plus intrinsic physical magnitudes of fields defined on that continuum in the nice "local" way captured in differential geometry. Lewis' vision of the HM was in fact very much like the familiar manifold + fields models of General Relativity and other so-called spacetime theories. Every point (or point + small neighborhood) is logically distinct and separable from every other, and all the happenings in the world are just the instantiations of local properties at these distinct places (or better, time-places).[2] In a Lewisian HM, all there is "one thing after another" and "one thing beside another."

[2] If we are to have anything more complex than scalar fields (such as a mass-density or charge-density fields)—that is, if we are to have vector fields or tensor fields (such as electromagnetic fields or metric fields) on our manifold, it strictly makes no sense to say that the physical events at a point are completely independent of those at any other point. The metrical and affine structure that must be encoded in tensors essentially

What was important for Lewis was that in the HM we find no trace of unwanted modality or connections between distinct existences, the sort of connections that Hume found so perplexing. Hume especially focused on *necessary* connections between distinct events in the HM, specifically, causal relations understood as amounting to event *A* making necessary the occurrence (elsewhere and/or later) of distinct event *B*. Lewis extended the puzzlement to two further notions: *laws of nature* understood in a necessitation-ascribing sense, and *chances* understood in a weakened-necessitation sense, such as one finds in some philosophical accounts of chance that appeal to chancy laws or chance propensities. For Lewis, then, the right way to approach making sense of such notions as *cause, law of nature*, and *objective chance* is by looking for philosophical theories that reduce facts about such things to some sorts of facts about the events in HM and how they are arranged over space and time. We will see how Lewis tried to do this for laws and chances in chapter 2.

My approach to objective chance will be different from Lewis', in that I will offer a Humean reductive account of chance, *but not of laws, causation, or anything else*. On these other topics I will mostly remain uncommitted in this book. Correspondingly, my HM specifically excludes only primitive chance-facts, and leaves as an open question whether there may be non-chancy sorts of necessitation between events in the HM, or things such as dispositions or powers (understood non-probabilistically). This is why I will avoid saying much more about what the HM corresponding to our world is like, and prefer to characterize the HM as I did in the preceding: it is simply everything that exists and the whole panoply of events that occur, in the history of our universe.[3]

---

determine how things change when one moves from a given point to its nearest neighbors in a given direction. Hence, such structured contents of the manifold can be defined "locally," but that means "at a point plus its immediate neighborhood" rather than "at a point" *simpliciter*. I take it that this much non-separability of events in the HM would not have troubled Lewis; the more radical sorts of connections between distinct spacetime regions that one finds in quantum theories are another matter entirely.

[3] In particular, I will not assume that our world is spatially 3-d as opposed to having some higher number of (spatial) dimensions. Later I will address the question of how to think about the temporal dimension of the HM.

1.2.2. Time and the HM

A classic Lewisian HM is a 4-d "block universe," to bring in a term from philosophy of time debates. That is, the HM includes everything that ever happens, anywhere, any*when*: future events as well as past events. There is no distinction of any significance whatsoever between the events in the HM that lie to our future versus those happening now or those that happened long ago. As we will see in later chapters, the fact that the HM includes future events as well as past ones brings both problems and benefits. But as long as we are taking care of metaphysical preliminaries, we should take some time to ask ourselves: do future events *really* "exist" in the same sense as present and past ones? Do we have a right to help ourselves to them as part of what grounds the facts (here, now) about objective chance?

At times, I confess, I have my doubts. Most of the time over my career I have assumed, and occasionally have defended, the B-series, block universe ontological perspective about time illustrated in Figure 1.1.

At the same time, I have also always felt the intuitive pull of the A-series perspective, according to which past and present events are quite distinct from (hypothetical, not yet real) future events. Suppose for a moment that the world is not governed by strictly deterministic

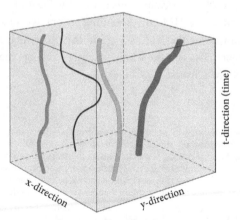

Figure 1.1. The block universe, with "things" existing over time.

laws, and that moreover whether there will be a sea battle tomorrow will be determined by the outcome of a photon polarization measurement on a photon prepared in a superposition state of the to-be-measured polarization. Given these supposed circumstances, I find it very plausible to agree with Aristotle that the proposition SB: <A sea battle occurs on [*tomorrow's date*]> is neither true nor false. There is not, at the moment, any fact in the universe that could ground the truth or falsity of this proposition. And this is to be understood not just in the sense captured by "There are no events at time [*insert current time*] which ground the truth or falsity of SB," but rather in the sense captured by "Quantifying unrestrictedly over everything that exists and every event in the whole universe's history, there are no events which ground the truth or falsity of SB." To put it more simply: the future is not real, does not exist, is not part of the world.

The reader may or may not feel any intuitive pull in favor of this A-series perspective. When I get into the state of feeling this pull, the perspective it leads me to corresponds, I think, not with any version of "presentism," but rather something more like a "growing block" perspective. The reason is simple: it seems to me that the world *does* contain all the facts one could wish to have, to ground the truth or falsity of propositions that are strictly about past events. The HM for our world is, in this sense, *at least* a growing block (GB) universe. For some thoughts about what this means, and how a GB universe should and should not be understood, see Box 1.2. But if our HM is (only) a GB, this means that the facts about objective chances, *as they are here and now,* depend only on the patterns of events in the past; and as new events are added to the GB, the patterns may of course change, and thus the objective chance facts themselves may evolve over time.

At appropriate places in later chapters we will come back to the question of how the Humean account of objective chance is affected by taking the HM to be a mere GB rather than a complete history of the universe. But for now, and the rest of this book (except where we explicitly take up issues concerning HOC in a GB), I will assume that the HM includes all future events as well as past and present ones.

We still need to consider a bit further what the *contents* of our HM block may be like.

Box 1.2 How to Think about a Growing Block World

In the literature on philosophy of time, the growing block model of temporal ontology tends to get a bad rap; it is widely taken to have all the disadvantages of both ordinary presentism and the "static" block universe, without having any offsetting advantages. I think this assessment is incorrect.

Let's set aside for the moment the stock objections to the very idea of time's passing—they have been ably refuted in (Maudlin, 2007a) chapter 4. Having set those aside, do we now have a clear and unproblematic understanding of what it is for time to *pass* or *flow*? Unfortunately, no, or at least, not an understanding that can be articulated clearly in other terms not obviously synonymous with *passage, flow*, etc. What about the 4-d block universe (BU) then, do we understand that better? Although I have tried most of the time to convince myself that I do, at times I cannot avoid the feeling that this is mere self-deception. There is really a sense in which it is fair to say that the BU is "static" and "unchanging." There is of course *also* a sense in which the BU contains things that change over time; without doubt, the BU has the resources to *model* time, and things changing over time. What it leaves out is anything corresponding to the *happening* of events, the unfolding of the world that turns (if the world is not deterministic, at least) manifold possibilities for the future into a single actuality. Time's passage must be taken to be a mere illusion of conscious experience, something that needs to be explained by the brain sciences— and explained using nothing more than what's found in the BU. There's the rub, in my view: every attempt I have seen to give an explanation of the illusion of time's passing seems to "cheat" a bit, telling a story that proceeds by tracing out processes unfolding over time, like a finger tracing this or that movement as it moves up a spacetime diagram on a page. In other words, the *actual* passing of time (in the telling of the story of what, in the block, allegedly explains the sense of time's passing) is relied upon in making the story seem somehow plausible.

To *that* kind of story, the A-theorist should not object at all, because she needs it, too! After all, even if time's passing is real, one does not (or should not) want to say that our experience of time's passing is some mysterious direct perception, unmediated by the movements of photons and the firings of neurons in our brains. No; our experience of time's passing is definitely something constructed in our brains, as many interesting temporal illusions illustrate (see Callender, 2017). But its construction is itself something that happens *as* time passes, and—the A-theorist should insist—at least in part *because* time passes, outside the mind, in physical reality.

Let's assume for now that we do want our understanding of time to include passage. Why model time with a GB rather than in one of the ways chosen by presentists? The answer is simple: while the future is not real and does not exist, the past *is* real—in the sense that all statements we can clearly formulate about past events are either true or false, not indeterminate, and are *made* true (/false) by the fact that the described events *did* (/*not*) *happen*. As Wittgenstein said, the world is everything that is the case, and that "everything" covers all past events, but not future events. If your view of time makes it somehow tricky for you to explain why "Dinosaurs roamed the earth millions of years ago" is *true*, you've taken a wrong turn somewhere. GB models of time, by contrast, capture the reality of the past (understood in this way) very clearly.

So far, so good. But the GB tends to get into trouble once the question is raised about the ontological *character* of the present, the "edge" of the block, vs. the parts that lie in the past. If the edge is taken to be no different from the rest of the GB, it seems we give ourselves an insoluble epistemic crisis: how do we know that we now, thinking these thoughts, are located at the moving edge of the block, rather than being stuck (like bugs in amber) somewhere back in the static part of the GB? How do we know the *now* is not several million years to the future of us? It's a pretty challenge, and a dangerous one too.[a] We may be tempted to say: Since *thinking* requires the passage of time, if we are now wondering to ourselves whether the now is *now*, that can only be because that thinking is in fact happening, i.e., is at the edge of the GB! But then we may be accused of populating the universe with zombies—all those near-copies of ourselves stuck in the past parts of GB, eternally "doing stuff" (it seems, if we trace the movements of particles upward through the block) are not in fact conscious beings at all! They are real (if the GB view is to be taken seriously) and look like us and "act" like us, but have no conscious minds—so they are zombies (to use the term of art from the philosophy of mind literature) (see Forrest, 2004).

Both the epistemic challenge and this unfortunate response to it are based on the mistake of thinking of the GB in the wrong way. If a 4-d BU does not really contain *time*, then neither does a truncated 4-d block. The right thing to say is: the world is a 3-d entity (*pace* string theories for now) that is constantly evolving and changing *as time passes*. So far, that sounds like presentism. And that's nearly right, but what the GB view adds is simply that the past, although ever-*growing*, is "fixed" in the sense that once an event has happened and become part of the past, it's there and there for good. That is, the past parts of the GB are part of the world in Wittgenstein's sense. But the physical world itself, this 3-d spread-out vast physical universe that keeps changing as time passes is best represented by the moving edge of the GB; the lower parts of the GB simply represent the facts about things that have happened in our world's history.

This, at least, is how I think a defender of genuine temporal passage should represent time's ontology. A huge further challenge arises if we try to square this ontological picture with relativity theory, and I will not take up that challenge here (interested readers should see (Correia & Rosenkranz, 2018) and (Pooley, 2013)).

[a] For discussions of this epistemic objection to the GB view, see (Braddon-Mitchell, 2004), (Miller, 2013), and (Correia & Rosenkranz, 2018, ch. 5).

1.2.3. Powers, Dispositions, etc.?

As we noted earlier, Lewis wished to defend a Humean perspective across the board. That means he wished to deny *any* form of modal connections between distinct events—so, no *laws of nature*—if these are understood as primitives or necessitation relations among facts or events; no *causation*—if this is taken as primitive, or as a necessitation relation; no *dispositions* (understood as primitives or as something like "causal powers"); and, to be sure, no *propensities* (understood as primitives or as *partial*-necessitation relations among facts or events). This desire to get rid of the modal led Lewis to insist that his HM involved only the existence at spacetime locations of "perfectly natural" properties, where these properties are *categorical* rather than dispositional. A categorical property is meant to be *there*, to really exist, but not to have as its nature something like a power or disposition or tendency. Consider mass density spread across a region of space, which we can represent using a scalar function ϕ that assigns a precise numerical mass-density property to each point of spacetime. (The function will assign value 0 to points where there is no mass.) There is a difference between a point where $\phi = 0$ and a point where $\phi = 53$, and also a difference between a point where $\phi = 53$ and a point where $\phi = 60$. These differences help make up the richness of that "whole panoply of events" that is the HM; and they can figure in regularities of a mathematical sort that may be grounded in HM, regularities which may be apt for being considered physical laws (or at least candidate-laws). But really, there is not much to such properties. They are simple, blah, inert, categorical properties. Charge density would be much the same as mass density, except that its values at points might include negative as well as positive real numbers. We would need to choose a different Greek letter such as ρ to represent it, and the patterns of variation of ρ's magnitude at different spacetime points might look different from the patterns of variation of ϕ. But that's about it, as far as differences between mass and charge would go.[4]

Could the HM of our world really be made up of nothing but spacetime locations instantiating purely categorical properties? The

---

[4] As we noted before, we could have more mathematically complicated fields such as tensors, vectors, and spinors representing occurrent properties in our HM. Such object fields would be better thought of as defined on infinitesimal regions of spacetime rather than on individual points, but this complication does not disrupt the essence of the Lewis-Humean picture of the HM.

idea of categorical properties seems both clear enough and harm-less enough at first sight; but when it is scrutinized a bit, doubts may emerge. In the first place, the properties we are used to seeing instantiated all around us seem to be individuated not merely by name, nor by their "patterns of instantiation" in spacetime, but rather by their *effects*, what they *do*. Or to put it in other language, most properties seem to be at the same time *powers* to affect other things, or *dispositions* to be affected by other things in certain ways. Consider the properties of *solidity* or *color*. The former is a power to resist the passage of other material things into the same space; the latter is a disposition to reflect certain frequencies of light more strongly than others, and hence a power to affect the visual system of animals like us in certain ways.

Coming back to the cases of mass density and charge density, intuitively what makes these properties different cannot be their names only, but rather something about their causal roles in na-ture. Mass density warps spacetime geometry without generating an electromagnetic field, and resists acceleration; charge density generates a field and attracts/repels oppositely/similarly charged objects; etc. *Prima facie*, our understanding of most properties that may be taken as things populating the HM is chock full of dispositionality and causality, and this has led some philosophers to question whether there really are any such things as purely cate-gorical properties.[5]

The Humean has a response to all these points, of course, which is not to deny the existence of cause-effect relations, dispositions, and so forth, but rather to insist that it is possible to give a reduc-tive story about all such modally involved notions, a reductive story that appeals only to the patterns of local property-instantiations in the HM. Of course it *seems* to us, and is handy to act as though, positively charged objects have the "power" to repel other positively charged things. But this can be understood as a fact reducible to the holding of certain laws of nature (something like Coulomb's law and/or Maxwell's laws of electromagnetism). And the truth of these laws,

[5] See for example (Sharon Ford, 2010).

in turn, can be given a reductive account in terms of a Humean Best Systems Analysis (BSA) of laws, as discussed in (Earman, 1986) and (D. Lewis, 1994).

I am sympathetic to the first part of this two-stage analysis of causes and dispositions, and skeptical about the viability of the second stage. Fortunately, for the purposes of this book I don't need to defend either of these stages. Readers who share Lewisian Humean attitudes may take the HM to be a mere palette intricately colored with inert categorical properties; dispositionalists and causal realists may take the HM to be governed by robustly modal laws of nature, or to be full of dispositionality and causality and what have you—*as long as these modal concepts are understood as not intrinsically **probabilistic**.* We will see later in this chapter why this is a reasonable restriction.

The goal of this book is to show how a Humean reductive account of objective probabilities supervening on patterns in our HM can give us everything we ever wanted from a theory of objective chance. Since the point is to reduce[6] facts about chances to facts about non-chancy features of the HM, I will simply be assuming that whatever the occupants of our HM may be like, they will not be "chancy" in and of themselves—as, say, *chance propensities* or *probabilistic dispositions* are presumably meant to be. In fact, I don't even need to assume that there are no such things in our HM. As Lewis said, "Be my guest—posit all the primitive unHumean whatnots you like. . . ." All I need ask of the reader is that she set those things aside for the moment, and consider my story about how Humean objective chances can be taken to reductively supervene on *everything else* in the HM.

That said, a reader might well wonder why she should bother finding out if some intricate Humean reductive program about

---

[6] Lewis' program of Humean Supervenience maintains, strictly speaking, that facts about laws, chances, causation, etc., *supervene* on facts about the HM, not that they can be *reduced* to such facts. In some contexts the difference between supervenience and reductionism may be significant, but here I believe it is not. Lewis provides explicit reductive stories, and my account of chance will do so as well. See (Dupré, 1993), chapter 4, for persuasive arguments to the effect that commitment to supervenience requires, in most instances, commitment to reductionism as well.

chance can work, if she is permitted to just posit primitive chance propensities as needed and be done with it? And the answer is that positing primitives is no panacea, especially when it comes to probability. It is hard, if not impossible, to see how one could find out about the values of primitive chance propensities, if such things existed; and if one somehow took oneself to know their values, it is still impossible to see why knowing them should in any way influence one's credences or one's actions. These problems, which taken together form the core of what I call the *dialectics of primitive chance*, make it hard to see that we even know what we are talking about if we posit chancy dispositions, powers, or propensities, or irreducibly chancy laws of nature. A proper appreciation of the problems will help the reader see why, for chance at least, a Humean reductive account is worth pursuing. The rest of this chapter will address these issues.

1.3. The Dialectics of Primitive Chance

> How and why should beliefs about objective chance help to shape our expectations of what will happen? This is the fundamental question about the concept of chance. I am going to argue that within the metaphysical point of view, the question cannot be answered at all.
>
> —van Fraassen (1989, p. 81)

Despite years of thinking about probability and chance, I've never been able to shake the feeling that I don't understand what is meant by those who postulate bare, unanalyzable, and underived primitive objective chances, or chancy fundamental laws. In this section my aim is to articulate my puzzlement as clearly as I can, and hopefully induce some readers to share it. And by exploring the dialectics of the debates between primitive chance advocates and skeptics, I will try to show that an alternative understanding of chance, one that *reduces* objective chance facts to facts of certain other kinds, is very much to be preferred. That alternative, Humean account will be developed and explored in chapters 2–5.

1.3.1. The Poverty of Primitives

Since the work of Peirce, at least in the early 1900s, there have been proposals that certain types of probability-talk be taken as primitive and basic, not to be cashed out or interpreted in terms of anything else. The idea is that there may be single events and/or types of events that are simply, irreducibly *chancy*: indeterministic, random, and only probabilistically predictable, i.e., associable with specific numerical chances between zero and one. Usually, when contrasted with other views about probability, this proposal is called the "propensity interpretation." But this is a misnomer, since the view is not an interpretation (= translation into other terms) so much as a flat-footed insistence that no interpretation in other terms can be given.[7]

While some philosophers find the propensity view[8] intuitive and relatively easy to understand, others accompany me in being puzzled about what could be meant by the postulation of primitive chances.

---

[7] I will use phrases like "the propensity view" or "primitive chances" or "fundamental chancy laws" interchangeably, as appropriate given the context. Some philosophers distinguish propensities from primitive chances, noting that the former has some conceptual link to the notion of *disposition*. I concede that likening propensity-chances to non-sure-fire dispositions may seem to be quite different from postulating a bare primitive probability and refusing to say more. But in the end, for reasons that hopefully will become clear in the following, I think this intended link actually adds nothing of content, so I will continue to treat propensity views and primitive chance views as one. The key question is: how do we understand the *numerical* part of a propensity statement? In the attempts to answer this question we will explore in the following, the possible moves are the same whether one makes the analogical connection to dispositions or not.

[8] The clearest example is to be found in (R. N. Giere, 1973); earlier papers by Popper are often cited but, in my opinion, they contain confusions about the locus of propensities and their connection to long-run frequencies. A variant of the propensity view close to Peirce's idea, and which does not *identify* propensities with objective probabilities, but still bases the latter on the former, is advocated in (Suárez, 2011) and (Suárez, 2016). (Nina Emery, 2015) mounts an inference to the best explanation argument for primitive chances; see Box 1.4.

Prominent among the puzzled are those who have Humean leanings, such as van Fraassen or David Lewis. In the famous passage mentioned earlier and which we will return to later, Lewis indicated his skepticism about primitive chances by labeling them "whatnots." But it may be that such skepticism is confined to a minority of contemporary philosophers; if so, the point of this section is to try to remedy that situation.

The basic question I want to pose and explore is this: what is the *meaning* of a putative primitive chance statement? What is the content of a statement such as $Pr_S(O) = .4$ (where S is an irreducibly chancy setup or situation, and O one of the possible outcomes), or of a putative chancy fundamental law statement such as $Pr_S(G_i) = x_i$ (where the G_i are the possible "outcomes" of the chance process covered by the law, and the x_i their objective chances)? What sort of claim are these statements supposed to be making about the world—and, in particular, what differentiates them from identical statements with numerically different x's on the right? How can we distinguish a possible world in which $Pr_S(O) = 0.4$ is true from one in which $Pr_S(O) = 0.55$ is true?

As we will see, it is not easy to answer these questions. In this section we will explore the dialectics of attempts to give answers, and then consider the arguments of those who say that no answer need be given.

The most powerful impetus for postulating primitive chances has come from the appearance in physics of an apparently fundamental and irreducible indeterminism, in quantum theories. Because of the successes of quantum theories, to many physicists and philosophers it now seems likely—or even definitively proven—that the laws of nature are irreducibly chancy.[9] So let's begin the dialectic by trying to say what the content or meaning of a (primitive, irreducible) chancy-law statement could be.

---

[9] Outside of the realm of quantum-derived chances, the main type of system to which we are wont to ascribe definite, precise objective chances, but unwilling to consider them to be mere frequencies, is classical gambling devices (coin flips, roulette wheels, card games, etc.). But most gambling devices can probably be modeled as deterministic systems with reliably randomly distributed initial conditions, and hence not genuine *loci* of primitive propensity-type chance, in the intended sense.

1.3.2.   Chance Laws?

On a traditional view of ordinary, non-probabilistic laws, one might say that the content of a law statement has two parts: an actuality-related part, and a modal part. The former part makes a claim about the way things are. In old-fashioned philosophy of science, we say that this part consists of a *universal generalization*—something that can be made to look something like "All F's are G's." Newton's law of gravity can be transformed into a statement of this form; the Schrödinger equation (which is not a probabilistic law—probability gets into the theory elsewhere, if it gets in at all) may or may not be expressable in this schema. But whether it can take this form or not, it still says something about actuality: the function Ψ representing a physical system with Hamiltonian H evolves over time according to

$$i\hbar\frac{\partial\Psi(q,t)}{\partial t} = \hat{H}\,\Psi(q,t).$$ Barring a strong anti-realist or instrumentalist stance towards Ψ, this law of quantum mechanics is making a definite claim about how things are (in this case, how they change over time) in the actual world. And the same goes for all other would-be fundamental laws that are not probabilistic.

The modal part of laws is perhaps not so clearly or uncontroversially understood, but is nevertheless taken to be an important part of the content. A law of nature, we want to say, does not simply say that such-and-so *does* happen, but moreover says that—in some appropriate sense—such-and-so *must* happen, i.e., that it is in some sense necessary.[10] But traditional accounts have never delivered a plausible and acceptable explication of this alleged necessity. Logical necessity seems too strong, to most philosophers. "Metaphysical" necessity strikes most philosophers as too strong to be the right modality also, and it has the additional drawback of being a controversial notion in its own right. Jumping to the far end of the spectrum, most

---

[10] To be more precise, it may not be the law itself (or a statement expressing the law) that says that such-and-so *must* happen; this may be correct to say for some accounts of laws and not for others. But it is correct to say that when *we* say that a certain generalization or equation is a law of nature, we intend to convey that it is not merely a truth about what does happen, but also tells us what *must* happen—again, in some appropriate sense of "must," which will vary from account to account.

philosophers are also not content to dismiss or internalize the modality, making it a mere matter of *our* attitudes and habits—though a long tradition beginning with Hume and continuing today pursues this option.[11]

So most philosophers who believe in laws of nature at all would like to have an intermediate-strength modality: *physical* necessity. How physical necessity should be understood is a thorny issue, which we need not explore. But it is worth noticing that philosophers and physicists at least make attempts, from time to time, to provide a deeper account of the nature and source of the physical necessity of laws. In recent analytic philosophy it has become fashionable again to argue that the necessity of physical laws is grounded in the essential natures of the basic physical kinds, and their powers or dispositions. Without an illuminating further account of these notions (essence, power, nature), of course, this is a relatively shallow account, and one that Humeans will reject. But it is something. As (Filomeno, 2014) discusses, there are much less shallow—though still highly speculative, and not yet successful—attempts at times by physicists to give a derivation of the necessity of the physical laws of our world. The point of mentioning such efforts is that they show that, when it comes to the modal aspect of laws, it seems at least conceivable that a substantial reductive account could be given, even if none has yet been achieved.

Whether or not any such program of explaining or deriving the necessity of laws of nature proves workable, though, philosophers will presumably not despair about non-chancy laws of nature, and for a perfectly good reason. At least *part* of the content of law claims seems clear: the part that says what actually is the case.[12] Now, what about chancy laws?

---

[11] Lewis' (1994) approach to laws, which we explore in chapter 2, eschews imputing any primitive modal status to laws, but also avoids internalizing the modality (or tries to do so at least). Lawhood is analyzed as membership in a "Best System," i.e., set of laws or axioms for our world (see later discussion for explanation). For Lewis, laws are then granted a kind of modal strength *via* their role in determining the truth values of various counterfactuals. Whatever its defects, the Lewisian approach leaves both the factual and the modal content of laws completely clear.

[12] This may help explain why many philosophers who are fans of laws but are unhappy with the modal part—e.g., Earman and Roberts (1999)—are deeply unhappy with the

We might expect that chancy laws also have two sides to their content: actuality-related and modal. Van Fraassen (1980, chapter 6) and others certainly think of objective probability as being a new, intermediate-grade sort of modality, intermediate between possibility and necessity, and equipped with its own logical rules, namely the probability calculus. So let's see if we can separate out the actuality-related and modal parts of the content of a probabilistic law.

Just as regular laws have a canonical logical form, "All F's are G's," probabilistic laws will have a canonical form too: $Pr(G_i|F) = x_i$, or perhaps $Pr_F(G_i) = x_i$. (The difference turns on whether we consider the setup conditions F to be part of the definition of the domain of the probability function, or conditions defined within a wider probability space, with the law specifying the probability of G conditional on F. We need not worry about the difference here.) The G_i are the possible "outcomes" of the chance process covered by the law, and the law gives us their objective chances.

Let's consider first the modal content of such laws. Do we understand it? It is not supposed to be mere possibility; true, we generally take it that any of the outcomes G_i that have non-zero chance are "possible," but that is not what we take to be the real story. The real modal part is supposed to have something to do with the number x_i, and to be "stronger" the closer x_i is to 1. What is this modality that comes in a continuum of relative strengths? Do we understand it by saying, say, "Since $Pr(G_1)$ is twice as large as $Pr(G_2)$, it is twice as possible"? No—this is meaningless. *Possibility* itself does not come in degrees— probability or "likelihood" does, but that is precisely what we are trying to analyze.

The modality here is usually, in a rough way, given a content in terms of a counterfactual—a statement about what *would* happen. The kind of thing we want to say goes like this: "If the F conditions were to be instantiated repeatedly, in a large number of causally independent 'trials,' then G_i would be the outcome in approximately an x_i fraction of the trials." We say this, but then we immediately take it back: "Of

---

idea that true, fundamental law statements should be read with a tacit *ceteris paribus* clause. If we undercut the contentfulness of the *actuality* side of law statements—which the addition of a *ceteris paribus* clause arguably does—what is left?

course, this need not always happen; it is simply *very likely* to occur, with higher and higher probability as the number of trials increases." But now it looks as though our counterfactual explication of the modal content of the law has failed; we have to demand an explication of the meaning of this *new* probability claim—both its actuality-related part (if it has any), and its modal part—and it is easy to see that we are off on an endless-regress goose chase. So the modal part of a chance law does not seem to have any clear content. We *want* to be able to cash it out in terms of something about what *would—definitely—*happen in certain circumstances, but we cannot. (Well, on the correct Humean account we *can*, as we'll see in chapter 3. But here we are just working from a traditional realistic take on chancy laws, i.e., the kind closest in spirit to the propensity account of objective chance.)

In Box 1.3, a different strategy for cashing out the modal content of chance statements is discussed, and similarly found wanting.

Box 1.3 A Further Attempt at the Modal Content of Chancy Laws

There is one more way to try to cash out the modal part of chancy laws' content, which has been explored in different ways by authors such as McCall (1994), Fetzer and Nute (1979), Giere (1976), and Pollock (1990). The idea is to analyze the objective chance as being some sort of *proportion*, among the branches representing possible futures of our world (McCall), among possible but non-actual "objects" or trials (Pollock), or among possible worlds (Giere, Fetzer). The idea seems promising at first: provide a non-empty semantic content for propensity ascriptions, one that at the same time makes clear the modal aspect of that content! Unfortunately, this rosy prospect endures only for as long as it takes to spell out one (any one) of these modal-proportion proposals.

Pollock proposed to identify $Pr(A|S) = x$ with <The proportion of physically possible S's that are A's is x>. We are to imagine a number of possible worlds, sharing the laws of our world, in which the relevant setup or conditions S are instantiated; and the proportion of those worlds in which A results is the objective probability we are trying to explicate. Based on Pollock's theory, we might say that the modal content of primitive chance statements is this: a fraction x of all (relevant) possible S- worlds are A- worlds. McCall, by contrast, proposes that we think of reality as having a branching-tree structure, where the past up to now is the trunk and the branches sprouting from now toward the future represent possible ways things can go. In the branching structure, there will generally be *multiple* A-branches and multiple branches for the other possible outcomes of the S setup; the objective chance of A in S is then identified with the proportion of A-branches out of the total number of branches starting with future chancy setup S. The diagram in Figure 1.2, adapted from (van Fraassen 1989, p. 84),

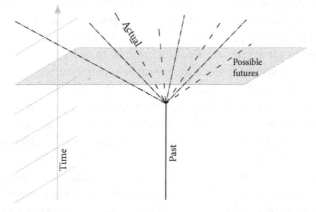

Figure 1.2. Chance may be defined in terms of the proportion of possible futures in which a certain outcome A obtains, out of the whole set of possibilities.

illustrates McCall's proposal. Fetzer and Giere also propose accounts that equate, in some sense or other, primitive chances with ratios or proportions or relative weights of sets of possible worlds.[a]

These explications of chance each have internal difficulties, some or all of which were discussed back in the 1970s when most of these works were done. But in addition, as a genre they face a few difficulties that seem to me insuperable.

In order to capture chances that take on any real value (i.e., irrational numbers such as $\frac{1}{\sqrt{2}}$ as well as rationals such as 1/3), the authors are forced to posit infinite structures (in some cases countable, in others continua). But ratios in infinite sets are notoriously problematic, being either undefined or requiring ad hoc assumptions or additions to the structure. Giere (1976), for example, ends up using continua in his semantic constructions, and positing ad hoc weight functions over sets of possible worlds.

But the worst problem that all these proposals run into is what van Fraassen called the "horizontal-vertical problem." What these proposals give us is a ratio when we "cut across" the relevant structure of possibilities horizontally, as in Figure 1.2. But the actual world is just one path through that upward branching tree, or one world in the whole set of possible worlds. In order for objective chances to justifiably guide our expectations in the way we want, we need to be sure they reflect, as appropriate, the *vertical* frequency of *A*'s in *S*-situations, the frequency found in the actual world. What reason do we have to think that the horizontal ratios or measures offered by these accounts are indicative of the vertical frequencies in the real world?

This last issue touches on a key aspect of the dialectics of primitive chances that we will explore further shortly. For now, though, we just need to take away the following upshot. Various authors have attempted to cash out the meaning of propensity-type chance claims in terms of ratios among sets of possible worlds or possible objects. But the elaborate mathematical constructions they created for this task all end up not being something we can take literally and seriously as ontology, and at best they serve as structures in which one can find a way to embed and represent the numerical chances that we started out wanting to understand. They do not fulfill the task of letting us understand the modal part of the content of primitive chance claims; and even if they did, it would remain a mystery why the chances so defined should tell us anything about what will happen in the actual world.

[a] Of these authors, although all four purport in some sense to be providing the *meaning* (or at least the "semantics") of propensity ascriptions, once the apparati have been presented only McCall seems to offer his in a serious vein, and about even his account I am not sure how literally we are meant to take it. Giere and Pollock both characterize their formal constructions as merely heuristic devices, and assure the reader that their framework of possible worlds/objects and measures over sets of them is not to be taken literally, but merely as a tool to *represent* propensity facts in a mathematically useful form.

[b] Giere essentially concedes the points I am making in his summary (p. 332), writing: "Metaphysically speaking, then, what are physical propensities? They are weights over physically possible final states of stochastic trials—weights that generate a probability distribution over sets of states. The function *u* provides only a formal and rather shallow analysis of this distribution. But is it not just the task of a propensity interpretation to explain what these weights are? No, because it cannot be done." The reader is left frankly puzzled about why an elaborate formal apparatus was needed, if in the end it is not to be taken seriously and does not shed any actual light on what it means to say that "*S* has a propensity of 0.*x* to give rise to *A*."

♠

Perhaps, as we found with non-chancy laws, at least the actuality-related content of a chancy law statement is clear? Unfortunately, it is not. Or at least, nothing capturing the numerical aspect appears to be in the offing. The actuality-related part cannot be merely the claim that whenever F are instantiated, one of the G_i results. (In fact, this is not necessarily so; the coin may, after all, land on its edge.) But we equally well know we cannot take the law to be saying that, *in actual cases where F is instantiated, a proportion* x_i *of the outcomes are* G_i. An actual frequentist, one kind of Humean about chance, can say this—but it is almost universally denied by philosophers today, and for good reasons. The actual frequencies may—in general, we expect, they *will*—diverge from the objective probabilities, especially when the number of actual cases in all of history is low. But if not this, then what is the actuality-related content of a chancy law? As far as I can see, *there is none*—nothing that is clear, true, and captures in some way the numerical-strength aspect of the chances. Ordinary laws of nature, on a straightforward realist approach, have a clear-cut actuality-related content, and a modal content that remains obscure. But chancy laws, on a straightforward realist approach, have no clear content of either sort.

In order to try to find the content of chance-law statements, I divided the search into two parts, one actuality-based and one modal; and we found it difficult to identify any clear content of either sort. But perhaps the problem is not with chance laws, but rather this proposed division of content itself? What happens if we go looking for the content of chance-giving laws *simpliciter*, without trying to split the modal from the actual/factual? I see no new options on the horizon. It seems that if we cannot offer a translation of (part or all of) the content of a propensity claim into something about frequencies of outcomes, we must instead stick to the notion of the *probability* of those outcomes. The only way to cash out the content in terms of probability without falling into obvious circularity is to understand "probability" in an epistemic sense, as a recommended degree of belief. We will return to this idea shortly.

1.3.3. Is a Primitive Kosher?

Now is the time to tackle head-on the likely primitivist response to the preceding: the whole search for a translation of the meaning of chance claims or chance laws in terms of *something else* was misguided. The whole point of positing objective chances or chance laws as a *primitive* is to make clear that it's not *possible* to give an explication in terms of something else. Nor is it needed; we (most of us!) understand perfectly well what is meant in making claims about objective chances, how to verify or falsify them, and how to use those claims. As Maudlin asks, "What more, exactly, could be wanted?"[13] With this we enter phase 2 of the dialectics.

Let's start with the primitivism espoused by Elliott Sober, which he calls the "No-theory theory" of probability. The basic idea of the

---

[13] (Maudlin, 2007b), "What Could Be Objective about Probabilities?"

METAPHYSICAL PRELIMINARIES 27

no-theory theory is that, in some cases at least, objective probability is not analyzable in other terms, and instead is a perfectly respectable theoretical concept. And like many important theoretical concepts, probability may not be definable in terms of other concepts, yet be perfectly well *implicitly* defined by its role in the theories that use it.

Sober (2010) offers us the analogy of the concept of (inertial) *mass* in physics. Despite the absence of a clean and clear definition, we believe that mass is an objective property of many physical objects, and for good reasons. For example, we have various different techniques for measuring or estimating mass, and they converge satisfactorily in most applications. The best explanation of this convergence is that it is due to a common cause, namely, that there *is* an objective property of mass (or rather, a certain quantity of mass) in the objects being measured. So even though we might consider that inertial mass is an unobservable property, perhaps even an intrinsically *dispositional* property, we can have perfectly respectable grounds for believing in its existence.

I find Sober's use of the theoretical concept of mass as an analogy for propensity-type probability unpersuasive. Mass is a property that, for a wide range of values, can be directly (if roughly) judged by human senses, and accurately measured with easily portable devices (e.g., scales). If a wide class of objective probabilities could be rapidly and accurately estimated by simple devices that we (say) insert the chance setups into,[14] then there would no doubt be a widespread and generalized sentiment among philosophers and scientists that the lack of an explicit theory was no serious obstacle to believing in the objective reality of probabilities.

But, of course, the real situation is quite different. There are no such devices. Moreover, there is no way to mount the sort of common-cause argument that Sober invokes as grounds for believing in the unobservable quantitative property of mass. When it comes to chances,

---

[14] There is a class of objective chance setups for which we seem able to judge the chances rapidly and accurately: classical gambling devices for which we feel we can deduce the chances on the basis of symmetry considerations. But this is a matter of inferring, not measuring; and as I will argue later, such systems are probably best understood as possessing objective probabilities grounded in determinism plus initial-condition facts, not primitive propensities.

we have exactly *one* way of measuring them: by looking at frequencies in (what we hope are) repeated instances of identical setups.[15] Everybody believes in mass and in its objectivity, even if a few peripheral applications are disputable. But among philosophers of probability, and scientists and statisticians who use it, there are substantial numbers who don't believe the concept picks out any *objective* reality at all (over and above actual frequencies). This sociological difference reflects the very real disanalogies between the concept of mass and the concept of primitive chance.

But if Sober's analogy is not quite convincing on its own, we still have to face the fact that many primitivists feel that they perfectly well understand the meaning of claims like "$Pr(G_i|S) = x_i$," and add that we understand both how to use them and how to confirm or disconfirm them. How could that be so, if my claim that we don't understand that primitive chance claims were correct?

1.3.4. Primitive Chances and the PP

The key to this part (phase 3) of the dialectics is the PP: the primitivists are relying on it, implicitly or explicitly, but the skeptic says they have no right to do so. What's happening in the mind of the primitivists who feel sure they understand the content of primitive chance claims is this: they translate the content of the chancy law into the obvious recommendations about what to expect (and how strongly), *via* an unconscious/automatic application of the PP.

---

[15] That said, with the advent of QM we have a new technique for *coming to know* certain objective chances: namely, deriving them from quantum wave functions, cross-sections, and so forth. (Something similar is true for probabilities given by classical statistical mechanics, but for most philosophers those are not considered candidates for being fundamental or primitive chances.) This novelty strengthens the grounds for considering QM chances to be objective and mind-independent, but does not at all resolve the puzzle about what propensity-type chance claims could *mean*.

As we saw earlier, the PP, in one simple form, says:

Let $Cr(\_|\_)$ be a rational subjective probability function (credence function), A be a proposition in the domain of the true objective chance function $P(\_)$, E be the rest of the agent's background knowledge, assumed to be "admissible" with respect to A, and X be the proposition stating that the objective probability $P(A)$ is x. Then:

$$Cr(A \mid XE) = x \qquad\qquad \text{(PP)}$$

Plugging chance-law probabilities into PP gives us the feeling that we understand the actuality-related content of the chance law: it is telling us that we *should* have such-and-so degrees of belief, so-and-such expectations for sets of repeated trials, etc. And we use these recommendations, often successfully; so there is our content, after all!

There is something deeply right about this, and we will see exactly what, how, and why in chapters 3–5. But as an explication of the *content* of probabilistic laws, understood in a non-skeptical, realist sense, it is a failure. Chance-credence principles like PP link objective probabilities to what it is rational to believe, and express an essential part of the nature of objective chance. But they are not, by themselves, an *analysis* of objective probability. PP does not tell us what objective probability *is*, it is simply a (candidate) constraint on rational belief. In order to be accepted as *in fact* a constraint on rational belief, it needs to be justified or demonstrated. And that demonstration would have to proceed *from* some facts about objective probability *to* facts about what rational agents' beliefs should be like. In other words, demonstrating or justifying the PP is a task for the philosophical analysis of objective chance (or chance laws) to fulfill. But as the quote from van Fraassen at the start of this section indicated, to say how and why beliefs about *primitive* chances should shape our expectations about what will happen appears to be impossible.[16]

---

[16] Furthermore, the recommendations that fall out of PP's application cannot be equated with the content of the original probability statements in question. Those recommendations are, after all, *value* claims, or perhaps imperatives, not factual assertions; they fall on the wrong side of the fact/value divide, assuming objective chance claims are intended to state facts.

Now is the time to come back to the famous quote from Lewis mentioned earlier. Referring to the possibility of positing intrinsically probabilistic dispositions, i.e., chance propensities, Lewis writes:

> Be my guest—posit all the primitive unHumean whatnots you like. (I only ask that your alleged truths should supervene on being). But play fair in naming your whatnots. Don't call any alleged feature of reality "chance" unless you've already shown that you have something, knowledge of which could constrain rational credence. (1994, p. 484)

The problem is that there is no way for the advocate of primitive chances to meet this challenge. Their alleged primitive chances, precisely by being bare primitives, lack any content beyond their numerical values. And being told that a system S has a whatnot-property G such that S's future possible behavior O has a numerical G-value x tells us *nothing whatsoever* that can or should clearly constrain our rational credences concerning O's occurrence or non-occurrence.[17]

To some philosophers it seems clear that the challenge here laid down to the advocate of primitive chances is unfair. Ned Hall responds to Lewis' challenge thus:

> What I "have" are *objective chances*, which, I assert, are simply an extra ingredient of metaphysical reality, not to be "reduced" to anything else. What's more, it is just a basic conceptual truth about chance that it is the sort of thing, knowledge of which constrains rational credence. Of course, it is a further question—one that certainly cannot be answered definitively from the armchair—whether nature provides anything that answers to our concept of chance. But we have long since learned to live with such modest sceptical worries. And at any rate, they are irrelevant to the

---

[17] Here I have used the letter "G" instead of "P" because we are so used to using "P" for "probability," and the legitimacy of that association is precisely what is at issue.
In case you feel that it is still somehow obvious or compelling that the G-values should be understood as chances, notice that nothing in the bare notion of G-values tells us to take them as probabilities *vs* taking them as 1–probabilities (i.e., G-values are one minus the probabilities), or any of a myriad of other possible interpretations.

question at issue, which is whether my primitivist position can "rationalize" the Principal Principle. The answer, manifestly, is that it can: for, again, it is part of my position that this principle states an analytic truth about chance and rational credence. (Hall, 2004, p. 106)

Hall's response to the challenge has, it seems to me, all the virtues of theft over honest toil. This form of argument is not one that we should be happy to extend to other areas of philosophical debate. Here is one close analogy, inspired by Strevens' paper on the PP (Strevens, 1999), which stressed the close ties between the problem of justifying PP and the problem of induction. Using Hall's strategy, one could solve the problem of induction in a trice:

What I "have" are *laws of nature*, which, I assert, are simply an extra ingredient of metaphysical reality, not to be "reduced" to anything else. What's more, it is just a basic conceptual truth about laws of nature that they are the sort of thing that hold in the future as well as the past. Of course, it is a further question—one that certainly cannot be answered definitively from the armchair—whether nature provides anything that answers to our concept of law of nature. But we have long since learned to live with such modest sceptical worries. And at any rate, they are irrelevant to the question at issue, which is whether my primitivist position can "solve the problem of induction." The answer, manifestly, is that it can: for, again, it is part of my position that it is an analytic truth about laws that they hold at all times.

Now, one may choose to *ignore* the problem of induction, and many law-primitivists probably do so. I would join them in this. But it is quite another thing to assert that the problem can be *solved*, by analytic *fiat*!

A more charitable reading of Hall's position about chances and the PP would be to consider that Hall builds in the connection in this way: objective chances are (by definition or stipulation) that feature of reality (if any exists) such that knowledge of it makes rational to have the corresponding credences as captured in PP. This strategy can be seen, for instance, in (Mellor, 1995). While it sidesteps the issue of justifying PP, this way of securing the validity of PP runs two risks. First, the risk that there is in fact no such feature, despite the evident existence of objective probabilities such as those of gambling devices and quantum theories. This is, however, a minor risk: if it strikes Hall down, then it strikes down everyone else equally. The real risk is rather that the definite description "that feature of reality such that knowledge of it makes rational to have the corresponding credences" may denote not some primitive dispositional entity or property, but instead the patterns in the HM that figure in a Humean reductive account! In the sequel we will see reasons to believe that precisely this is in fact the case.

Box 1.4 An IBE to Primitive Chances?

In a pair of recent papers on chance (Emery, 2015, 2017), Nina Emery discusses the claim that chances (which could be "nomological," primitive/*sui generis*, or propensities) may be posited in order to explain facts about frequencies for certain types of events. Her 2015 discussion centers on a "paradigm case" of objective chance, the up-deflection probability of silver atoms passed through a Stern-Gerlach device. The silver atoms in Emery's paradigm case have been prepared in a deflected-up state along axis x by previous measurement interaction, and subsequently are measured along axis x' which is not far from x; so the quantum probability of upward deflection is high, and indeed most atoms are deflected up. Emery stipulates that no hidden variable differentiates the prepared silver atoms in a way that could explain specific outcomes; this stipulation is part of what makes it her "paradigm case." Turning to explanation, Emery writes:

[There is a] more general claim which, given the observations she has made, is liable to be part of Sally's best theory about the experimental setup:

(5) The chance of any silver atom in the experiment being deflected up is very high.

And (5) demonstrates that in addition to claims about frequency providing evidence for claims about chance, claims about chance can also, at least sometimes, explain claims about frequencies. In particular, (5) explains (6):

(6) Most of the silver atoms in the experiment have been deflected up.

That (5) explains (6) may not be immediately obvious, but the argument for it is straightforward. First, notice that if (5) does not explain (6), then nothing does. It is part of the paradigm case that there is no feature such that all and only silver atoms that are deflected up have that feature before they are sent through the magnets. What else, then, could explain the fact that most of the silver atoms sent through the experiment are deflected up? What other sort of fact could be provided as answer to the question, "Why have most of the silver atoms in the experiment been deflected up?" Second, notice that if nothing explains (6), then we have no reason for expecting the pattern described in (6) to continue. If it is just a massive coincidence that most of the silver atoms sent through the experiment have been deflected up, then we should not expect the next silver atom to be deflected up. But, we do expect the next silver atom to be deflected up, and most of the silver atoms after that. So we expect the pattern described in (6) to continue. So something explains (6). And the only possible explanation is (5).

(2015, p. 113)

There are a number of ways to resist Emery's argument and its conclusion. Some empiricists (though not I) would dispute the claim that if we believe nothing explains the existence of a certain regularity, then we have no reason to expect it to persist. Moreover, it is controversial whether primitive chances do provide an explanation of stable probabilistic regularities. Emery's explanans and explanandum are worded fairly vaguely, presumably deliberately, so as to accommodate the potential gap between the chance and the observed actual frequency. But the wiggle room between primitive chances and frequencies, which in principle has no bounds, can be as much foe as friend to the explanation-based defense of primitive chance. For example, suppose that the actual frequency behind Emery's (6) is 97%, but the quantum chance is supposed to be only 89%. If that frequency is built up over many thousands of silver atoms, then standard statistical practice would be to regard this frequency as strongly disconfirming, if not outright refuting, the hypothesis that the objective chance is in fact 89%. As the frequency gets further and further away from the putative objective chance, *intuitively*, the alleged explanation of the frequency by the chance gets worse and worse, eventually becoming no explanation at all; but the semantic content of a primitive chance claim gives us no clue about how to justify or quantify this intuitive claim. This brings us back to the considerations of section 1.3, which can be read as an extended argument that primitive chance claims cannot explain actual frequencies or our expectations about them, because (i) we don't know what the semantic content of primitive chance claims could be, and (ii) we have no justification for plugging them into PP to generate subjective expectations.

The reader may or may not agree with these objections. But there is a further question to ask, before we sign on to an IBE argument in favor of primitive chances: Even if primitive chances provide *a* possible explanation for stable probabilistic regularities, do they give us the *best* explanation? In Hoefer (2016) I argued that they do not: deterministic underpinnings give an even better explanation. We will see an example of this, for coin flip frequencies, in chapter 3. Emery excludes the possibility of such underpinnings in her example by stipulation, but as Werndl (2011) (and the example of Bohmian mechanics) shows, it is at least difficult and perhaps impossible to exclude determinism at the fundamental dynamical level.

In her 2017 paper Emery argues that the frequency-explanation issue is a *prima facie* problem for a Humean approach to chance: Since, for Humeans, frequency-facts at least partly ground or determine the chance facts, the latter cannot then be taken to *explain* the former, on pain of a vicious circularity. This issue of explanation is certainly an important one, and we will return to it in later chapters. For now, it may suffice to note that

Humeans about chance are ready to accept that chances do not explain frequencies *in the way that powers or dispositions* (allegedly) *explain events, or the way necessitarian laws of nature* (allegedly) *explain regularities.* There is nonetheless a different sort of explanation that Humeans will lay claim to: nomic subsumption. As Hempel would have said: given the chance fact (5), the truth of (6) was *to be expected.* In so far as a Humean account of chance and *only* a Humean account is able to justify the Principal Principle (see chapter 4), one might even argue that a Humean approach to chance does *better* on this score than does a primitive or propensity approach!

1.4. Chance as Hypothetical Infinite-Run Frequency: Propensities Made Respectable?

In their (1993), Howson and Urbach offer a similar diagnosis of the problems of a propensity account of chance. Concerning Giere's (1973) account, they write:

> The second objection is more fundamental and seems to be unanswerable. Von Mises's theory may seem stubbornly deficient in empirical content, but the present account is, if anything, even worse. For Giere's single-case propensity theory conveys no information of any sort about observable phenomena, not even, we are told, about relative frequencies in the limit. It is not even demonstrable, without the frequency link, that these single-case probabilities are probabilities in the sense of the probability calculus . . . nor do the limit theorems of the probability calculus help. (pp. 341–342)

But Howson and Urbach (1993) go on to argue that there *is* an account of objective chances that is based on the notion of propensities (in a sense), for which all these problems can be resolved. What they defend is a form of *hypothetical frequentism,* specifically von Mises's theory of probabilities as long-run frequencies in collectives.

While hypothetical frequentism is sometimes presented as an analysis of objective probability that is completely different from propensity accounts, it is possible to view it as a one way of fleshing out what Gillies (2000) calls a "long-run propensity view" (though not the same as Gillies' own version of such a view). As Howson and Urbach (1993) note, this is clear in (von Mises, 1957/1981). Objective chances are associated with repeatable "experiments" (chance setups), which are assumed to have a tendency or disposition to give rise to certain stable limit frequencies of the possible outcomes, in the long run. After this acknowledgment of the link to the notion of propensity or disposition, however, these terms do no further work. The real content comes in through von Mises's notion of a *collective*.

A collective is an infinite sequence of (hypothetical) experiment outcomes which has two key properties:

1. Each possible outcome, as well as finite Boolean combinations of the possible outcomes, has a well-defined limiting frequency; the convergence is well-behaved, there being for every such outcome and every $\varepsilon > 0$ a finite number N in the sequence after which the frequency of the outcome in the finite partial sequence of outcomes from 1 to $M > N$ always stays within ε of the infinite limiting frequency;
2. No infinite subsequence generated by a recursively definable place-selection rule has limiting frequencies, for any outcome(s), that are different from those in the full collective.

Property 1 ensures that our hypothetical infinite sequences are well-behaved, while property 2 ensures that they are "random" in a desirable sense.[18] The metaphysical content of the von Mises theory of chance, also endorsed by Howson and Urbach, is then simply the posit that there are physical systems in the world which are such that, if they

---

[18] This is so except in the extreme where the entire sequence consists of the same outcome (e.g., an infinite sequence of heads outcomes for coin flips). Such a sequence meets both of von Mises' conditions, but arguably is intuitively as far from random as can be. As long as the frequency is strictly between zero and one, however, this issue does not arise.

were to be "run" an infinite number of times, the outcomes they produce would constitute a von Mises collective.[19] Which systems in the world have these genuine-chance dispositions is a matter for empirical science to decide, though we can presumably be confident that well-designed classical gambling systems do fit the bill.

The traditional objection to any hypothetical frequentism is a simple, but overwhelmingly important one: if objective chances are defined as limits in hypothetical or counterfactual infinite sequences of trials, it seems clear that nothing *we* can ever observe in this world can give us any clue about what the objective chances are.[20] As Howson and Urbach put it,

> The very serious, because apparently justified, charge against von Mises's theory is precisely that the ideal entities it postulates *cannot* be incorporated into scientifically successful theories: the theory seems to be irremediably metaphysical. On a limiting relative-frequency interpretation of probability statements, *a hypothesis of the form* $P(h) = p$ *makes no empirically verifiable or falsifiable prediction at all*, for it is well known that a statement about the limit of a sequence of trials hypothetically continued to infinity contains by itself absolutely no information about any initial segment of that sequence. (1993, p. 331)

Howson and Urbach point out that all of von Mises's attempts to overcome this objection were unsuccessful. But chance statements, understood as claims about hypothetical frequencies, *would* be testable if we could demonstrate the validity of the PP for them. (The same is true of propensity claims in general, for that matter). The idea would be to give a Bayesian treatment of our credences about the objective chances, and use Bayes' rule to calculate how our beliefs narrow in on a particular hypothesis (or small range of hypotheses) about the

---

[19] We can reasonably demand, of the proponent of the von Mises analysis, that she give us a story about what, in reality, makes the counterfactuals she invokes true; and we should demand that the answer not involve hand-waving talk of propensities or primitive dispositions. But I will set this concern aside.

[20] I.e., what the values of the chances, such as $Pr(heads|flip)$, are. But we can also add: all our evidence, necessarily finite, will be wholly unable to tell us which physical systems even *have* associated objective chances, and which ones do not.

objective chances, as evidence accumulates.[21] PP plays the key role in this procedure: it takes us from the objective likelihood $Pr(e|H_p)$ of getting evidence (outcomes, results of trials) e conditional on the true chances being those specified in H_p to matching subjective credences. With the subjective likelihoods in hand, we can use conditionalization to update our credences about the chances as evidence accumulates, and in favorable circumstances we will narrow in on a single hypothesis H_p and become effectively "certain" of it in a reasonable time. This, arguably, responds to the worrisome objection that objective probability claims are entirely untestable.

The critical link in this line of reasoning is the justification of PP, so we must look carefully to see whether Howson and Urbach give us a convincing demonstration that PP is valid for chances-as-von-Mises-hypothetical-frequencies. Their argument for PP is an a priori argument, with resemblance both to the Dutch Book argument for coherence of rational credences and to the ordinary-language argument for the rationality of induction. Here is how it goes:

> The use of the Principal Principle is justified as follows. Suppose that you are asked to state your degree of belief in a toss of this coin landing heads, conditional upon the information *only* that were the tosses continued indefinitely, the outcomes would constitute a von Mises collective, with probability of heads equal to p. And suppose you were to answer by naming some number p' not equal to p. Then, according to the definition of degree of belief [credence] in Chapter 5, you believe that there would be no advantage to either side of a bet on heads at that toss at odds $p':1-p'$. But that toss was specified *only* as a member of the collective characterised by its limit-value p. Hence you have implicitly committed yourself to the assertion that the fair odds on heads occurring at *any* such bet, conditional just on the same information that they are members of a collective with probability parameter p, are $p':1-p'$. Suppose that in n such tosses there were m at which a head occurred; then for fixed

stake S (which may be positive or negative) the net gain in betting at those odds would be $mS(1-p') - (n-m)Sp' = nS[(m/n) - p']$. Hence, since by assumption the limit of (m/n) is p, and p differs from p', you can infer that the odds you have stated would lead to a loss (or gain) after some finite time, and one which would continue thereafter. Thus you have in effect contradicted your own assumption that the odds $p':1-p'$ are fair. Thus we have established a version of the Principal Principle . . ." (Howson & Urbach, 1993, p. 345)

This is a beautiful argument, and one that has not been properly appreciated in the chance literature.[22] At bottom it is an argument about coherence among one's beliefs, in the sense of *logical consistency*, not probabilistic coherence. It shows that *if* you conceive of objective chances as von Mises did, *and* you take yourself to know the von Mises chance p of an outcome A in a chance setup, then you must set your credence in A to p for any trial of the chance setup, on pain of logical inconsistency.[23] Does it succeed in helping us overcome the epistemic disconnect between finite world evidence and hypothetical infinite sequences?

The argument has, it seems to me, one weak point. There are physical/metaphysical difficulties entrained by the infinitude of a von Mises collective, that make it questionable whether one *could* have the kinds of beliefs that occur in the premises of Howson and Urbach's argument, unless one is ignorant of quite basic physics. *This* coin, that is about to be tossed, cannot—as a matter of physical possibility—be tossed an indefinite number of times; it will wear out sufficiently to cease to have a coin-like shape after only a finite number of flips. So (recognizing this) one could not coherently believe *also* that if the coin were to be flipped indefinitely, a von Mises collective of outcomes

---

[22] Both (Strevens, 1999) and (B. Loewer, 2001) mischaracterize the Howson and Urbach argument as being "consequentialist" (i.e., an argument that one will do best to adhere to the PP), which it is not.

[23] Notice that to derive an inconsistency, our fictional agent must accept Howson and Urbach's definition of credences (degrees of belief) as belief-about-fair-betting-quotient. Eriksson and Hajek (2007) argue that all definitions of subjective credence based on the notion of betting are failures, but I think Howson and Urbach's definition resists their criticisms (see Box 1.1).

would result. Perhaps we can rephrase our belief as one about a collection of coins, each starting off with identical physical properties, and a collective being generated out of the flips of these coins—worn out ones being replaced as needed? But infinity strikes us down again: the collection of coins would have to be infinite, as would the time required to do the flipping. So to have a coherent counterfactual belief about infinite coin flips, we must assume the infinity of the mass content of the universe *and* the infinity of future time.[24] But in any world recognizably like ours (i.e., originating in a Big Bang), either time is finite or, if infinite, the eternal expansion plus thermodynamic laws entails that all processes (including coin flips) grind to a halt in a finite time.

So a physically educated person *cannot* have the belief that the next toss of a certain coin is part of a hypothetical von Mises collective; even as a counterfactual assumption set on the right-hand side of the conditionalization bar, von Mises chances require one to abandon very basic beliefs about physics, if contradictions are to be avoided.

This may seem like unfair nit-picking, this taking seriously the physics of a hypothetical infinity of flips. The von Mises collective is always put forward, after all, as an idealization, like point masses in Newtonian physics.[25] Very good, but let's be careful as we relax into thinking of von Mises chances as mere idealizations: we need (a) to keep clearly in view *what they are* (since they are meant to be objective facts), and we need (b) to check whether the Howson and Urbach justification of PP can go through if von Mises chances are taken as something else, merely idealized as an infinite collective.

[24] Perhaps we could reconstitute previously used-up coins by scraping the copper off of the surfaces of our flipping room. Still, that takes energy, so the following point in the text makes this irrelevant. If we try to avoid an infinite future by having infinite coins being flipped at different spatial locations over a finite span of time, then as Hájek (2008) points out, the limiting frequency becomes dependent on how the spatially separated flips are ordered into a sequence. In any case, it is far from clear that we can stretch our belief about what happens when *this coin* is tossed repeatedly into some fact about hypothetical infinite collections of coexisting simultaneous flips spread out over infinite space.

[25] See Alan Hájek's "Fifteen Arguments against Hypothetical Frequentism" (2009) for compelling criticisms of the claim that seeing von Mises collectives as merely idealizations somehow saves the analysis.

Points (a) and (b) are difficult to satisfy jointly. (a): Once we embrace the idealized character of von Mises chances explicitly, the most honest translation of the objective content of a chance statement would seem to be something like this: " 'Pr(H|flip) = p' means that if a *very* large (but finite) number of flips were to be performed under relevantly similar conditions, then the results would (with high probability) be a frequency of H near to p, with a random-looking outcome sequence." This, as we will see, is very close to what my Humean skeptical account of chances says, for coin flips (deleting the "(with high probability)"). But it cannot be what the hypothetical frequentists mean to say, since they rely on the infinitude of collectives to get most of the juicy mathematical theorems they prove about probabilities and chances. Von Mises himself, for example, explicitly denies that we can make sense of randomness in finite outcome sequences (1957, p. 83). Moreover, (b) Howson and Urbach's justification of PP will not work if the hypothetical collective of coin flips is finite in extent. Though "unlikely," the frequency *(m/n)* might never get closer to p than it is to p', in a finite sequence of flips, so our fictional agent whose credence is set to p' is not incoherent. But if my suggested de-idealization of the von Mises chance claim is not the correct one, what is? I see no way to make the Howson and Urbach justification of PP work without the agent's beliefs involving an infinite sequence of flips, where the infinity is taken seriously (albeit as hypothetical); but I also see no way for a realistic agent to take these infinitistic claims seriously, even hypothetically.[26]

Howson and Urbach's argument aims to show that if you believe that the coming coin tosses have a von Mises chance of H and that its value is p, then you are incoherent if you don't set your credence in H equal to p. If it worked unproblematically, this would be a great result to have proven. But there would still be more that one might hope to be able to say in favor of PP: that adherence to it is, *in results*, a better

---

[26] Perhaps there is some other way to de-idealize von Mises chance statements, without dropping the infinity; but I see no way to make this work that does not at the same time undermine the alleged incoherence used in Howson and Urbach's deduction of PP. In chapter 4 I will modify the Howson and Urbach argument for use in the service of Humean objective chance; infinity will not have to be invoked in order to derive a contradiction from failure to respect PP.

strategy than non-adherence. The Howson and Urbach argument does not show that, if the von Mises chance *does* exist and *is p*, then an agent who bets on a large number of tosses is going to do better if his credence in H is set to *p* than he/she will if it is set significantly differently. This sort of justification of PP can only be given for Humean chances, as we will see in chapter 4. Objective chance is meant to be a "guide to life," and I take that to mean *our lives*, in all their unfortunate finitude. It would be nice to know, not just that we *must believe* chance to be a good guide, but also that it *really is* a good guide to life.

1.5. Summing Up

In this chapter we have had first contact with a number of metaphysical notions that play important roles in discussions about objective probability and chance, in particular: *time, laws of nature, causation, disposition*.

The foundation of the Humean view of chance to be elaborated in this book is the Humean Mosaic: all the events there are in the history of the world. The HM provides the factual subvenience basis on which the facts about objective chance will supervene. We looked at two ways in which the temporal aspect of the HM might be understood: in a "block universe" manner (facts about events to our future are included), or in a "growing block" manner (only facts about present and past events exist in the required sense to subvene other facts). Except where explicitly noted, in the rest of this book we will understand the HM in the block universe way, so that facts about chance will supervene on future as well as past patterns of events.

When it comes to dispositions and laws of nature, I will remain as neutral as possible on the question of how we should understand them, and indeed on the question of whether there are such things at all. The only assumption that I will ask the reader to make with me is that there are no *primitive, probabilistic* dispositions (propensities) or laws of nature (chance laws). I've tried to justify this request in the preceding by throwing doubt on whether we really understand what claims that there are such things could mean.

Finally, what about causation, which has so far been mentioned only in passing? Well, one way that chance propensities are sometimes characterized is as non-surefire causes of a particular sort, namely, ones to which we ascribe a definite numerical chance-value. This sort of causation, clearly, I am asking the reader to set aside as not clearly meaningful. But otherwise, as with laws, I will try not to take any particular stance concerning how we ought to think about causation, if there is such a thing in the world. This will be true even in chapter 8, which will come back to the connections between probability and causation. There I will argue that the real link between causation and probability is on the *epistemic*, or credence side of probability: knowing that a cause is in place gives reason to raise one's credence in the occurrence of the effect. On the ontic side, are there connections between objective chances and causation? I will argue that there are not.

The main lessons I hope the reader takes away from this chapter are the following. Propensity accounts of chance postulate the existence of something objective, and further claim that this something is apt for guiding our credences; but there is no reason to believe in these magical things unless we are given a substantive story about *what* they are and *why* they should guide credence. We entertained the possibility that if we, in effect, build our chance propensities into the basic laws of nature, things might go better. But—unsurprisingly, in hindsight—that did not help matters. Without a substantive analysis in terms of something else, it is unclear that we can say that chancy laws have any content at all. At least, we found, there is no way to spell out either a factual (actuality-based) or a modal content for such laws. Defining chances as hypothetical infinite-limit frequencies did not help, despite Howson and Urbach's ingenious attempt to justify the PP for hypothetical-frequency chances.

This situation, in which no story about what objective chances *are* seems to work, is unacceptable. What sort of account of chance can get us out of this situation? One that, ideally,

- entails that statements about objective chances have a clear factual content;

- entails that such statements have a clear modal content (even if that is achieved by denying that any such content exists); and
- lets us see why plugging the chances into the PP is indeed reasonable.

These goals will be met in the sequel. Before diving into correct views about chance, however, readers may wish to look at Box 1.5 for a brief discussion of a popular viewpoint on probability that denies the need for any analysis or theory of objective chance.

Box 1.5 On Subjectivism

"Probability does not exist!" declaimed Bruno de Finetti at the beginning of his *Theory of Probability* (de Finetti, 1990), and legions of philosophers and scientists have taken his message to heart. These *subjectivists* (or *radical subjectivists*, or *Bayesian personalists*) believe that *objective* probabilities do not exist, and that everything we do with probabilities in science and in daily life can—indeed, *must*—be understood as involving subjective degrees of belief, i.e., credences. Whether it be the probability of rain tomorrow, the probability of getting snake-eyes when rolling a pair of dice on a craps table, or the probability of a radium atom's decaying in the span of one year, all we are *really* talking about are credences, and nothing in the world exists that directly makes certain credences correct and others incorrect. All probabilities are simply psychological states.

What motivates the subjectivists to take this hard line against objective probabilities? In essence, just the sort of problems we have been discussing throughout chapter 1, as well as the familiar difficulties that refute older analyses (classical and logical accounts, finite frequentism—see A. Hájek, 2008). Their attitude is not unreasonable: if in centuries of development of the theory and application of probability, not one philosophically acceptable analysis has been able to explain what probabilities *are* in a way that shows them to exist as objective facts "outside the head," perhaps it is time to take the hint and accept that probabilities only really exist inside the head. The work of Ramsey (1926) (reprinted in Eagle, 2010) and de Finetti (1937) laid the foundations for subjectivism by providing an operational definition of credences in terms of betting quotients and showing how the probability axioms can be justified on grounds of practical rationality, using the famous Dutch Book arguments.

Despite all the achievements of the Bayesian subjectivists, it is hard to accept that if two craps players in Las Vegas have very different probabilities for snake-eyes—one having $\frac{1}{36}$, the other $\frac{1}{2}$—there is no objective sense in which the former is correct and the latter mistaken. It is even harder, I think, to accept the subjectivist line when it comes to quantum mechanical probabilities. Of course, it is not as though the subjectivists have *nothing* to offer in place of objectivity for chances. They can make use of (actual) frequencies to serve as a surrogate for objective probabilities—for example, noting that the credence of our $\frac{1}{36}$ gambler is more in tune with the frequency of snake-eyes on craps tables, and that this a useful property for his credence to have (and also one that he is sure to arrive at, if he updates his credences using Bayes' theorem *after* starting from a reasonable set of "priors").

But as we will see in later chapters, frequencies are not a good enough surrogate for the probabilities of quantum mechanics. Nor do they help us with the nagging feeling that, in the case of snake-eyes on craps tables, it's not just a *happenstance* that the frequency, over all history, is very close to $\frac{1}{36}$. The Humean account to be offered in chapter 3 can be thought of as offering a sophisticated package of amendments to actual frequentism. It will address that nagging feeling, and handle quantum mechanics gracefully; but none of the amendments invokes the kind of metaphysical or *a prioristic* elements that de Finetti and other subjectivists reject. The chief motive for embracing subjectivism (or Sober's no-theory theory) has been the (correctly) perceived failures of past objectivist analyses. Chapters 3–5 aim to eliminate this motive. Moreover, the theory of chance to be offered will share much of the spirit, and some of the letter, of both the subjectivist's and no-theory theorist's viewpoints. So there is hope for an outcome of peaceful coexistence!

2

From Lewisian Chance
to Humean Chance

Because Lewis' approach to objective chance is well known, in this chapter I will introduce his view, and work toward my own Humean view by amending Lewis' at several important places.[1]

2.1. Principal Principle (PP)

One of the two shared fundamentals of Lewis' account and mine is the claim that the Principal Principle (PP) tells us most of what we know about objective chance. Recall that PP can be written:

$$Cr(A \mid XE) = x$$

where X is the proposition that the objective chance of A is x, and E is the rest of one's evidence, presumed "admissible."[2]

The idea contained in PP, an utterly compelling idea, is this: if all you know about whether A will occur or not is that A has some objective probability x, you ought to set your own degree of belief in A's occurrence to x. Whatever else we may say about objective chance, it

---

[1] For a clear exposition and defense of Lewis' approach, see (Barry Loewer, 2004).

[2] In Lewis' presentations of PP, $Cr(\_)$ is presented as a rational *initial* credence function (what Alan Hájek likes to call a "Superbaby" credence function), and E is any proposition, not necessarily a proposition representing an agent's evidence. But my sketch here is faithful to the way that Lewis actually applies and uses the PP, especially in (D. Lewis, 1994). The difference between Superbaby credence functions and credence functions representing real agents such as ourselves can be important in some contexts, but the distinction can be safely ignored for our purposes here.

has to be able to play the PP role. PP captures, in essence, what objective chances *are for*, why we want to know them.[3]

Crucial to the reasonableness of PP is the limitation of E to admissible information. Suppose you are contemplating the first quantum spin measurement you will perform in your lab next week. Normally it would make sense to set your credences in the possible outcomes according to the objective chances, which you take quantum theory to have revealed. But if you happen to have a friend whom you believe to be a(n honest) time traveler from the future, and she has already told you that the first spin measurement next week will yield the result *up*, evidently you are not rationally obliged to set your credences to the quantum chances. Your background information E contains inadmissible information.

What makes a proposition admissible or non-admissible? Lewis defined admissibility completely and correctly in 1980, I believe—though he considered this merely a vague, first-approximation definition:

> Admissible propositions are the sort of information whose impact on credence about outcomes comes entirely by way of credence about the chances of those outcomes. (1986a, p. 92)

This is almost exactly right. When is it rational to make one's subjective credence in A exactly equal to (what one takes to be) the objective chance of A? When one simply has *no information* tending to make it reasonable to think A true or false, except by way of making it reasonable to think that the objective chance of A has a certain value. If E has any such indirect information about A, i.e., information relevant to the objective chance of A, such information is canceled out by X, since X gives A's objective chance outright. Here is a slightly more precise definition of

---

[3] This point applies to any sort of objective probability as well (for those who prefer to distinguish several sorts of objective probability and call only one of them "chance"). Such non-subjective, non-purely-epistemic probabilities are automatically trusted to guide our expectations, plans, and previsions, which is to say we automatically apply PP to them.

> Admissibility: Propositions that are admissible with respect to outcome-specifying propositions A_i, for an agent whose credence function is $Cr(\underline{\quad})$, contain only the sort of information whose impact on the agent's credence about outcomes A_i, $Cr(A_i)$, if any, comes entirely by way of impact on the agent's credence about the chances of those outcomes.

This definition of admissibility is clearly consonant with PP's expression of what chance is for, namely guiding credence when no better guide is available. The admissibility clause in PP is there precisely to exclude the presence of any such "better guide."

Three things are worth mentioning about this definition. Lewis' glosses on admissibility all implicitly treat it as though it were an impersonal relation between propositions. In fact, since we are dealing with credence functions, i.e., personal subjective probabilities, what we are talking about is not "impact" in an objective/impersonal sense, but impact-for-the-agent. If someone's personal credence function treats E as irrelevant to the truth of A, where E is some sort of information that most people regard as *highly* relevant to A's truth, then E is nevertheless admissible information for that agent.[4]

A second point to notice is that admissibility is relative to the proposition A whose subjective probability is in question. In 1980 and 1986a Lewis seems to have favored instead a definition of admissibility that is not agent- or proposition-relative, but rather *time*-relative only. As we will see, this may have led Lewis astray in some regards; it was corrected by Lewis and Hall in their separate (1994) papers.

---

[4] There are niceties and difficulties that can be raised here, which we will not for the moment go into. For example, if the information E is some sort of information about actual frequencies, and it *does* have an impact on credence in A for our agent, how can we tell whether that information has its impact *via* credence about chance, vs. some more direct route? Second, how can we tell when E does have impact on credence about A? Is it sufficient that $Cr(A|EK)$ and $Cr(A|-EK)$ be the same? What if E is in fact taken to be causally relevant to A, but *via* two causal paths, one positive and one negative, and the agent simply happens to feel (at the moment) that these paths evidentially cancel each other? For most of the uses we make of admissibility in this book, intuitive judgments of admissibility will be clear-cut enough to get by with. I regard it as an open question whether an acceptable and formally precise definition is possible or not.

Third, it is worth noting that this definition does not make the PP tautologous. An agent could have no inadmissible information (as defined in **Admissibility**) and yet still have conditional credences that fail to satisfy PP. **Admissibility** also does not say that it is *sufficient* for E to be admissible that $Cr(A|XE) = Cr(A|X)$ (see note 4 for one reason why this is not a consequence of the definition.)

In my definition of admissibility, there is no mention of past or future, complete histories of the world at a given time, or any of the other apparatus developed in Lewis (1980/1986a) as (what he thought were) first steps toward a correct and precise definition. We will look at some of that apparatus in the following, but it is important to stress here that none of it is needed to properly understand admissibility. Lewis' substitution of a precise sufficient condition for admissibility in place of the correct definition seems to be behind two important aspects of his view of objective chance that I will reject in the following: first, the alleged "time-dependence" of objective chance; second, the alleged incompatibility of chance and determinism.[5]

2.2. Time and Chance

As we just noted, Lewis did not feel that his informal characterization of admissibility was precise enough. In order to get at least a precise *sufficient* condition for admissibility, and cast chance-credence issues in a framework he found congenial, Lewis introduces some time- and world-indexing and a special case of PP, which we will call:

$$Cr\left(A \mid H_{tw}T_w\right) = x = Pr_t(A) \qquad \text{(PP2)}$$

In this formula, H_{tw} represents the complete history of the world w up to time t, and T_w represents the "complete theory of chance for

---

[5] Lewis' belief that a different characterization of admissibility was needed caused other problems as well. For example, in the context of his "reformulated" PP, which we will see later, it caused Lewis to believe for a long time that the true objective chances in a world had to be *necessary*, i.e., never to have had any chance of not being the case. This misconception delayed his achievement of his final view by well over a decade.

world w." T_w is a vast collection of "history-to-chance conditionals." A history-to-chance conditional has as antecedent a proposition like H_{tw}, specifying the history of world w up to time t; and as consequent, a proposition like X, stating what the objective chance of some proposition A is. The entire collection of the true history-to-chance conditionals is T_w, and is what Lewis calls the "theory of chance" for world w. $Pr_t(\_\_)$ is the objective chance function at time t. As Lewis conceives of chance, it is a time-dependent notion; plug the entire history of the world up to t into T_w, and it will then tell you the probability, at t, of A being true. To the extent that we can think of Lewis as positing what other authors call "chance setups," setups are just initial segments of total world-history. As we will see, an awful lot is being taken on board by recasting the chance-credence relationship in terms of these tools!

Lewis was a trenchant opponent of metaphysical propensity accounts of chance. And his Humean Supervenience doctrine is presented in terms of a "block universe" viewpoint. So it is ironic that Lewis claims, with the propensity theorists, that the past is "no longer chancy." If A is the proposition that the coin I flipped at noon yesterday lands heads, then the objective chance of A is *now* either zero or one—depending on how the coin landed. (It landed tails, so on Lewis' view $Pr_{now}(A) = 0$.) Unless one is committed to an A-series view such as the growing block, and the associated view that the past is "fixed" whereas the future is "open," there is little reason to make chance facts time-dependent in this way. I prefer the following way of speaking: my coin flip at noon yesterday was an instance of a chance setup with two possible outcomes, each having a definite objective chance. It was a chance event. The chance of heads was $\frac{1}{2}$. So $\frac{1}{2}$ is the objective chance of A in that setup. It still is; the coin flip is and always was a chance event. Being to the past of me-now does not alter that fact, though as it happens I now know A is false.

[What if one *does* commit to a growing block world, as I confessed I have on occasion been tempted to do? In that case, the objective chances of events are liable to change as time passes—not due to their location in history, but rather because T_w itself may change as the HM, the basis on which T_w supervenes, grows larger and more events are added to the world's history. But one may still consider yesterday's coin flip to be a chancy event with an intermediate chance between zero and one.]

PP, with admissibility properly understood, is perfectly compatible with taking chance as not time-dependent. It seems at first incompatible, because of the sufficient condition for admissibility Lewis gives, which says that at any given time t, any historical proposition—i.e., proposition about matters of fact at or before t—is admissible. Now, a day after the flip, that would make $\neg A$ itself admissible; and of course $Cr(A|\neg AE)$ had better be zero (1980/1986a, p. 98). But clearly this violates the correct definition of admissibility. $\neg A$ carries maximal information as to A's truth, and not by way of any information about A's objective chance; so it is inadmissible. My credence about A is now unrelated to its objective chance, because I know that A is false. But as Ned Hall (1994) notes, this has nothing intrinsically to do with time. If I had a reliable crystal ball, my credences about some *future* chance events might similarly be disconnected from what I take their chances to be. (Suppose my crystal ball shows me that the next flip of my lucky coin will land "heads." Then my credence in the proposition that it lands "tails" will of course be zero, or close to it.)

There is a real asymmetry in the amount and quality of information we have about the past, versus the future. We tend to have lots of inadmissible information about past chance events, very little (if any) inadmissible information about future chance events. But there need be nothing asymmetric or time-dependent in the chance events themselves.[6] Taking PP as the guide to objective chance illustrates this nicely. Suppose you want to wager with me, and I propose we wager about yesterday's coin toss, which I did myself, recording the outcome on a slip of paper. I tell you the coin was fair, and you believe me. Then your credences should be $\frac{1}{2}$ for both A and $\neg A$, and it's perfectly rational for you to bet either

---

[6] There *need be* no time asymmetry to objective chances, but often there *is* a presupposed time-directedness. Typically chance setups involve a temporal asymmetry, the "outcome" occurring after the "setup" conditions are instantiated. But in no case do the categories of past, present, or future (as opposed to before/after) need to be specified.

way, with even stakes. (It would only be irrational for you to let *me* choose which way the bet goes.) The point is just this: if you have no inadmissible information about whether or not *A*, but you do know *A*'s objective chance, then your credence should be equal to that chance—whether *A* is a past or future event. Lewis (1980/ 1986a) derives the same conclusions about what you should believe, using the PP on his time-dependent chances in a roundabout way. I simply suggest we avoid the detour.[7]

2.3. Lewis' Best System Analysis (BSA) of Laws and Chance

Lewis applies his Humeanism about all things modal across the board: counterfactuals, causality, laws, and chance all are analyzed as results of the vast pattern of actual events in the world.[8] This program goes under the name "Humean Supervenience," or HS for short. Fortunately, we can set aside Lewis' treatments of causation and counterfactuals here. But his analysis of laws of nature must be briefly described, as he explicitly derives objective chances and laws together as part of a single "package deal":

Take all deductive systems whose theorems are true. Some are simpler, better systematized than others. Some are stronger, more informative, than others. These virtues compete: an uninformative system can be very simple, an unsystematized compendium of miscellaneous information can be very informative. The best system is the one that strikes as good a balance as truth will allow between simplicity and strength. . . . A regularity is a law *iff* it is a theorem of the best system. (1994, p. 478)

---

[7] By avoiding the detour, we also avoid potential pitfalls with backward-looking chances, such as are utilized in Humphreys' objection to propensity theories of chance (see Humphreys, 2004).

[8] For Lewis (unlike me) that HM pattern is simply the pattern of instantiation of "perfectly natural properties" throughout space-time. Lewis' commitment to antinominalism at the level of fundamental physical properties may be seen as one modal commitment that he does not try to recover by means of analysis in other terms.

Lewis modifies this BSA account of laws to make it able to incorporate probabilistic laws:

> we modify the best-system analysis to make it deliver the chances and the laws that govern them in one package deal. Consider deductive systems that pertain not only to what happens in history, but also to what the chances are of various outcomes in various situations—for instance, the decay probabilities for atoms of various isotopes. Require these systems to be true in what they say about history. We cannot yet require them to be true in what they say about chance, because we have yet to say what chance means; our systems are as yet not fully interpreted. . . .
>
> As before, some systems will be simpler than others. Almost as before, some will be stronger than others: some will say either what will happen or what the chances will be when situations of a certain kind arise, whereas others will fall silent both about the outcomes and about the chances. And further, some will fit the actual course of history better than others. That is, the chance of that course of history will be higher according to some systems than according to others. . . .
>
> The virtues of simplicity, strength and fit trade off. The best system is the system that gets the best balance of all three. As before, the laws are those regularities that are theorems of the best system. But now some of the laws are probabilistic. So now we can analyze chance: the chances are what the probabilistic laws of the best system say they are. (1994, p. 480)

A crucial point of this approach, which makes it different from actual frequentism, is that considerations of symmetry, simplicity, and so on can make it the case that (a) there are objective chances for events that occur seldom, or even never; and (b) the objective chances may sometimes diverge from the actual frequencies even when the actual "reference class" concerned is fairly numerous, for reasons of simplicity, fit of the chance law with other laws of the System, and so on. My account will preserve this aspect of Lewis' Best Systems approach. Law facts and other sorts of facts, whether supervenient on Lewis' HS-basis or not, may, together with some aspects of the HS-basis "pattern" in the events of the world, make it the case that certain objective chances exist, even if those chances are not grounded in that pattern alone. Examples of this will be discussed in chapter 3.

Analyzing laws and chance together as Lewis does has at least one very unfortunate consequence. If this is the right account of objective chances, then *there are* objective chances only if the best system for our world says there are. But we are in no position to know whether this is in fact the case, or not; and it's not clear that further progress in science will substantially improve our epistemic position on this point. Just to take one reason for this, to be discussed further later: the Lewisian best system in our world, for all we now know, may well be deterministic, and hence (at first blush) need no probabilistic laws at all.[9] If that is the case, then on Lewis' view, contrary to what we think, there aren't any non-trivial objective chances in the world at all.

This is in my view a very undesirable feature of Lewis' account. Objective probabilities *do* exist; they exist in lotteries, in gambling devices and card games; and possibly even in my rate of success at catching the 9:37 train to work every weekday. In science, they occur in the statistical data generated in many physical experiments, in radioactive decay, and perhaps in thermodynamic approaches to equilibrium (e.g., the ice melting in your cocktail). If a view of chance leaves open the whole question of whether any non-trivial chances exist (i.e., chances strictly between 0 and 1), then the notion of "objective chance" captured by that view is not the notion at work in actual science and in everyday life.

It is understandable that some philosophers who favor a propensity view should hold this view that we don't know, and may never

---

[9] Lewis points to the success of quantum mechanics as some reason to think that probabilistic laws are likely to hold in our world. But a fully deterministic version of quantum mechanics exists, namely Bohmian mechanics. (Suppes, 1993) offers general arguments for the conclusion that we may never be able to determine whether nature follows deterministic or stochastic laws; see (Werndl, 2011) for a more in-depth treatment of this issue.

know, whether there are such things as objective chances (though it is, I believe, a serious bullet to have to bite). It is less clear why Lewis does so. On the face of it, it *is* a consequence of his "package deal" strategy: chances are whatever the BSA laws governing chance say, which is something we may never be able to know. But if we (as I urge) set aside the question of the nature of laws, and think of the core point of Lewis' Humean approach to chance, it is just this: objective chances are simply facts following from the vast pattern of events that comprise the history of this world. *Some* of the chances to be discerned in this pattern may in fact be consequences of natural laws; but why should *all* of them be?

Thinking of the phenomena we take as representative of objective chance, the following path suggests itself. There may be some probabilistic laws of nature; we may even have discovered some already. But there are also *other* sources of objective chances that probably do not follow from laws of nature (BSA or otherwise): probabilities of drawing a Queen in a card game, of getting lung cancer if one smokes heavily, of being struck by lightning in Florida, and so on. Only a very strong reductionist would think that such probabilities must somehow be *derivable* from the true physical laws of our world, if they are to be genuinely objective probabilities; so only a strong reductionist bias could lead us to reject such chances if they *cannot* be so derived. And why not accept them? The overall pattern of actual events in the world can be such as to make these chances exist, whether or not they deserve to be written in the Book of Laws, and whether or not they logically follow *from* the Book. As we will see in chapters 3 and 4, they are there because they are capable of playing the objective chance-role given to us in the PP.

Suppose we do accept such objective chances not (necessarily) derivable from natural laws. That is, we accept non-lawlike, but still objective, chances, because they simply are *there* to be discerned in the mosaic of actual events (as, for Lewis, the laws of nature themselves are). Let's suppose then that Lewis could accept these further non-lawlike chances *alongside* the chances (if any) dictated by the Best System's probabilistic laws. Now we can turn to the question of whether objective chances exist if determinism is true.

2.4. Chance and Determinism

Lewis considers determinism and the existence of non-trivial objective chances to be incompatible. I believe this is a mistake.[10]

In 1986a Lewis discussed this issue, responding to Isaac Levi's charge (with which I am, of course, in sympathy) that it is a pressing issue to say how to reconcile determinism with objective chances (Levi, 1983). In his discussion of this issue (1980/1986a, pp. 117–121) Lewis does not *prove* this incompatibility. Rather he seems to take it as obvious that, if determinism is true, then all propositions about event outcomes have probability zero or one, which then excludes non-trivial chances. How might the argument go? We need to use Lewis' sufficient condition for admissibility and his revised formulation of PP,

$$(PP2) \quad C\left(A \mid H_{tw} T_w\right) = x = Pr(A)$$

in which H_{tw} represents the complete history of the world w up to time t, and T_w represents the "complete theory of chance for world w." T_w is a vast collection of "history-to-chance conditionals." A history-to-chance conditional has as antecedent a proposition like H_{tw}, specifying the history of world w up to time t; and as consequent, a proposition like X, stating what the objective chance of some proposition A is. The entire collection of the true history-to-chance conditionals is T_w, and is what Lewis calls the "theory of chance" for world w. Suppose that L_w are the laws of world w, and that we take them to be admissible. Now we can derive the incompatibility of chances with determinism from this application of PP2:

$$C\left(A \mid H_{tw} T_w L_w\right) = x = Pr\left(A\right)$$

For determinism is precisely the determination of the whole future of the world from its past up to a given time (H_{tw}) and the laws of nature

---

[10] In the years since Lewis' (1994) many philosophers have joined me in arguing that non-trivial objective chances, or objective probabilities (for those who distinguish these terms), must be compatible with determinism. See for example (Loewer, 2001), (Glynn, 2010), and (Lyon, 2010).

(L_w). But if H_{tw} and L_w together *entail* A (say), then by the axioms, $Cr(A|H_{tw}T_wL_w)$ must be equal to 1 (and *mutatis mutandis*, zero if they entail (-A)). A contradiction can only be avoided if all propositions A have chances of zero or one. Thus PP2 seems to tell us that non-trivial chances are incompatible with deterministic laws of nature.

But this derivation is spurious; there is a violation of the correct understanding of admissibility going on here. For if $H_{tw}L_w$ entails A, then it has a *big* (maximal) amount of information pertinent as to whether A, and not by containing information about A's objective chance! $H_{tw}L_wT_w$ may entail that A has chance 1. That's beside the point; if it's a case of normal deterministic entailment, $H_{tw}L_wT_w$ also entails A itself (using just $H_{tw}L_w$). And that is carrying information relevant to the truth of A other than by carrying information about A's objective chance. So $H_{tw}L_w$, so understood, must be held inadmissible, and the derivation of a contradiction fails.

PP, properly understood, does not tell us that chance and determinism are incompatible. But there is another way of thinking of Lewis' approach to chance that may explain his assumption that they are incompatible, already alluded to earlier. It has to do with the "package deal" about laws. Lewis may have thought that deterministic laws are automatically as strong as strong can be; hence if there is a deterministic best system, it can't possibly have any probabilistic laws in its mix. For they would only detract from the system's simplicity without adding to its already maxed-out strength.

If this is the reason Lewis maintained the incompatibility, then again I think it is a mistake. Deterministic laws may not after all be the last word in strength—it depends how strength is defined in detail. Deterministic laws say, in one sense, almost nothing about what actually happens in the world. They need initial and boundary conditions in order to entail anything about actual events. But are such conditions to form part of Lewis' axiomatic systems? If they can count as part of the axioms, do they increase the complexity of the system infinitely, or by just one "proposition," or some amount in between? Lewis' explication does not answer these questions, and intuition does not seem to supply a ready answer either. What I urge is this: it is not at all obvious that the strength of a deterministic system is intrinsically maximal

and hence cannot be increased by the addition of further probabilistic laws. If this is allowed, then determinism and non-trivial objective chances are not, after all, incompatible in Lewis' system. Nor, of
course, will they be incompatible on the account I offer.

2.5. Chance and Credence

Lewis (1980/1986a) claims to prove that objective chance is a species of probability, i.e., follows the axioms of probability theory, in
virtue of the fact that PP equates chances with certain ideal subjective
credences, and it is known that such ideal credences obey the axioms
of probability.

> A reasonable initial credence function is, among other things,
> a probability distribution: a non-negative, normalized, finitely
> additive measure. It obeys the laws of mathematical probability
> theory. . . . Whatever comes by conditionalizing from a proba-
> bility distribution is itself a probability distribution. Therefore
> a chance distribution is a probability distribution. (1980/
> 1986a, p. 98)

This is one of the main claims of the earlier paper justifying the title "A
Subjectivist's Guide. . . ." But it seems to me this claim must be treated
carefully. First, ideal rational degrees of belief are shown to obey the
probability calculus only by the Dutch book argument, and this argument seems to me only sufficient to establish a *ceteris paribus* or
prima facie constraint on rational degrees of belief. The Dutch book
argument shows that an ideal rational agent *with no reasons to have
degrees of belief violating the axioms* (and hence, *arguably*, no reason
not to accept any wagers valued in accord with her credences) is irrational if she nevertheless does have credences that violate the axioms.
By no means does it show that there can never be a reason for an ideal
agent to have credences violating the axioms. Much less does it show
that finite, non-ideal agents such as ourselves can have no reasons for
credences violating the axioms. Given this weak reading of the force of
the Dutch book argument, then, it looks like a slender basis on which

to base the requirement that objective probabilities should satisfy the axioms.[11]

Lewis' chances obey the axioms of probability just in case T_w makes them do so. It's true that, given the role chances are supposed to play in determining credences *via* PP, they ought *prima facie* to obey the axioms. But there are other reasons for them to do so as well. Here is one: the chances have, in most cases, to be close to the actual frequencies (again, in order to be able to play the PP role—see chapter 4), and actual frequencies are guaranteed to obey the axioms of probability.[12] So while it is true in a broad sense that objective chances must obey the axioms of probability because of their intrinsic connection with subjective credences, it is an oversimplification to say simply that objective chances must obey the axioms because PP equates them with (certain sorts of) ideal credences, and ideal credences must obey the axioms.

Second, on either Lewis' or my approach to chance, it's not really the case that objective chances are "objectified subjective credences" as Lewis (1980/1986a) claims. This phrase makes it sound as though one starts with subjective credences, does something to them to remove the subjectivity (according to Lewis: conditionalizing on $H_{tw}T_w$), and what is left then plays the role of objective chance. In his reformulation of the PP, Lewis presents the principle as if it were a universal generalization over all reasonable initial credence functions (RICs):

Let C be any reasonable initial credence function. Then for any time t, world w, and proposition A in the domain of P_{tw}

$$[\textbf{PP2}:] \qquad P_{tw}(A) = C(A \mid H_{tw}T_w)$$

In words: the chance distribution at a time and a world comes from any reasonable initial credence function by conditionalizing on the complete history of the world up to the time, together with the complete theory of chance for the world. (1980/1986a, pp. 97–98).

---

[11] There are other arguments in the credence literature that show, under this or that further assumption, that credences must satisfy the axioms of probability. I see all of them as offering at most *ceteris paribus* or *conditional* reasons for conforming to the axioms. (If you want your credences to have such-and-such properties, then . . .).

[12] Setting aside worries that may arise when the actual outcome classes are infinite.

Read literally, as a universal generalization, this claim is just false. There are some RICs for which the equation given holds, and some for which it does not, and that is that. It is no part of Lewis' earlier definition of what it is for an initial credence function to be reasonable, that it must respect PP![13] But, clearly, any RIC that does not conform to PP will fail to set credences in accordance with the preceding equation.

PP is of course meant to be a principle of rationality, and so perhaps we *should* build conformity to it into our definition of the "reasonable" in RIC. This may well be what Lewis had in mind.[14] Then Lewis' claim in the preceding becomes true by definition. Nevertheless, the impression it conveys, that somehow the *source* of objective chances is to be found in RICs, remains misleading. The source is T_w, which—for Humeans—is in turn based on the Humean Mosaic *via* the Best System Analysis of chance.

It is misleading in a second way as well. The RIC function's domain presumably covers all, or nearly all, propositions; so $C(A|H_{tw}T_w)$ is a quantity that can be presumed to exist, in general, for any A. But Lewis intended in the "Subjectivist's Guide" (1980/1986a) to be cautious about the domain of objective chance, and *not* to presuppose that it is defined over all propositions (pp. 131–132). Lewis does restrict PP2 to propositions A in the domain of the chance function in the preceding quote. But elsewhere he, and most of those who follow his approach, tend to forget this restriction and treat the domain of objective chance as essentially coextensive with the domain of the credence function. This has the effect of obscuring what should be viewed as two very live possibilities: (a) that the chance function for the actual world is patchy and non-comprehensive; and (b) that even if micro-physical chances exist and are, in some sense, "complete," we reject the strong reductionism that would take those micro-chances to entail chances for any conceivable macro-level proposition. These points will be discussed in detail in chapter 5.

Humean objective chances are simply a result of the overall pattern of events in the world, an aspect of that pattern guaranteed, as we will

---

[13] RICs are usually understood to be credence functions that obey the probability axioms, assign credence 1 only to logical truths and credence 0 only to logical contradictions, and perhaps are structured so as to permit "learning from experience" in a world such as ours.

[14] See (1980/1986a), pp. 110–111.

see, to be useful to rational agents in the way embodied in PP. But they do not start out as credences; they *determine* what may count as "reasonable" credences, via PP. In Lewis' later treatments this is especially clear. The overall history of the world gives rise to one true "Theory of chance" T_w for the world, and this theory says what the objective chances are wherever they exist.

2.6. Summing Up

We have taken on board two crucial theses from Lewis' account of objective chance: that the chance facts should supervene on the overall mosaic of actual events in the full history of the world (the full block universe); and that whatever objective chances are, they must be *demonstrably* able to play the role ascribed to them in PP, of serving as a guide to credence in situations of (partial) ignorance. Several other points of Lewis' theory we have found in need of correction, and the Humean approach to chance that has begun to emerge can be seen already to have these advantages:

1. It allows us to be sure that non-trivial chances exist, even though (for all we know) our universe may be subject to deterministic laws of nature;
2. It allows us to treat past and future symmetrically, even though it may turn out that most chances of interest to us are temporally "forward-looking";
3. By disentangling discovery of chances from discovery of the true laws of nature, it lets us keep chances within our epistemic grasp—subject only to the usual and unavoidable doubts that attend all inductive inference.

After we lay out the positive story of Humean objective chance in chapter 3, we will be able (in chapter 4) to demonstrate what Lewis only claimed to "see dimly, but well enough," namely the justifiability of plugging Humean chances into the Principal Principle.

3
Humean Objective Chance

So far I have been laying out the Humean view of chance indirectly, by correcting a series of (what I see as) mistakes in Lewis' treatment. Now let me give a preliminary, but direct, statement of the view I advocate. This approach has much in common with Lewis' as amended in the preceding—but without the implied reductionism to the microphysical. I will lay out this preliminary statement one piece at a time, each piece being a sentence in italics accompanied by some discussion.

3.1. The Basic Features

Chances are in the first instance probabilities of outcomes, in the context of the instantiation of a proper chance setup, and additionally such probabilities as can be derived from the basic chances with the help of logic and the probability axioms.

An initial question to resolve is whether to take the basic form of an objective chance statement to have the form of a *conditional* probability, or an *unconditional* probability: $P(A|B)$, or simply $P(A)$?

Originally, I followed Alan Hájek (2003) in considering *conditional* chance as the more basic notion, rather than unconditional chance. Hájek argues for this on a variety of grounds, mostly based on criticisms that he levels against the canonical Kolmogorov "definition" of conditional probability in terms of unconditional probability:

$$Pr(A|B) = Pr(A \& B) / Pr(B)$$

This formula makes conditional probability undefined when $Pr(B) = 0$, but there are intuitively cases where $Pr(B) = 0$ and yet the conditional probability seems both well-defined and obvious.

One of Hájek's favorite examples (slightly modified): given a chance setup in which an infinitely sharp dart is thrown "at random" onto a perfectly circular disk, *Pr(lands on left side|lands on the horizontal center-line)* $= \frac{1}{2}$. Despite this and other considerations, I am now more inclined to stick with unconditional (but setup-relative!) probability and the usual Kolmogorov axioms, for my theory of objective chance.

When it comes to paradigmatic examples of objective chance, if they are to be conditional probabilities, it seems to me that what we would want to put on the right hand side of the conditionalization bar is the chance setup. In terms of propositions, the conditioning proposition would specify what chance setup is instantiated in the "trial" or "run" or "experiment" at issue. I will usually denote the chance setup with letter S, so if we were to take conditional probability as the basic notion, then the basic form of an objective probability would be the conditional "$Pr(A|S)$," where A is one of the possible "outcomes" of the setup S.

But we sometimes will want to put other things on the right-hand side of the conditionalization, as in things like *Pr*(die lands 6|die is rolled *and the result is even*) $= \frac{1}{3}$. Such a conditional probability would be written with this form:

$$Pr(A|S \& B) = \frac{1}{3}$$

The trouble with this is that it suggests that S and B are in some sense on a par; and with the help of Bayes' theorem we can calculate $P(S|A\&B)$. Mathematically there is no problem here, $P(S|A\&B) = 1$, which makes sense given the way we are thinking of S: as the background context in which the objective chances exist. On the other hand, it may tempt us to think of the occurrence of S (e.g., "a fair die is rolled") as, itself, a chancy phenomenon, and we do *not* want to presume that this is the case. As Hájek reminds us, the probability that I get heads in the context of a fair coin flip is $\frac{1}{2}$; but the probability that I flip the coin? That may not even exist. So I prefer to keep S as a subscript denoting the chance setup, rather than treat it as proposition in the domain of an objective probability function. Thus, the canonical form of an objective chance statement throughout this book will be "$P_S(A) = x$," and unless otherwise noted, Kolmogorov's axioms with finite additivity will be presupposed.

Chances are constituted by the existence of patterns in the mosaic of events in the world. The patterns have nothing (essentially) to do with time or the past/future distinction, and also nothing to do with the nature of laws or determinism. Therefore, neither does objective chance. Still, in most—but not all—of the examples of interest to us, it is the case that the instantiation of the chance setup occurs before the instantiation of the "outcome" event.

These patterns are such as to make the adoption of credences identical to the chances rational in the absence of better information, in a sense to be explored in the following. Sometimes the chances are just finite/actual frequencies; sometimes they are an idealization or model that "fits" the pattern, but which may not make the chances strictly equal to the actual frequencies. (This idea of "fit" will be explored through examples, in the following and in chapter 5).

It appears to be a fact about actual events in our world that, at many levels of scale (but especially micro-scale), events look "stochastic" or "random," with a certain stable distribution over time; this fact is crucial to the grounding of many objective chances. I call this the Stochasticity Postulate (SP). We rely on the truth of SP in medicine, engineering, and especially in physics. The point of saying that events "look stochastic" or "look random," rather than saying they *are* stochastic or random, is dual. First, I want to make clear that I am referring here to "product" randomness, not "process" randomness. Sequences of outcomes, numbers, and so on can look random even though they are generated by (say) a random-number-generating computer program whose input-output function is deterministic. For the purposes of our Humean approach to chance, looking random is what matters. Second, randomness in the sense intended is a notion that has resisted perfect analysis, and is especially difficult when one deals with not-huge finite sequences of outcomes. Nevertheless, we all know roughly how to distinguish a random-looking from a non-random-looking sequence, if the number of elements is high enough. Our concern at root, of course, is with the applicability of PP. Sets or sequences of events that are random-looking with a stable distribution will be such that, if forced to make predictions or bets about as-yet-unobserved parts of them (e.g., the next ten tosses of a fair coin), we can do no better than adjust our expectations in accord with the objective chance distribution.[1]

Some stable, macrocscopic chances that supervene on the overall pattern are explicable as regularities guaranteed by the structure of the assumed chance setup plus SP. These cases will be dubbed *Stochastic Nomological Machines* (SNMs), in an extension of Nancy Cartwright's (1999) notion of a nomological machine. A nomological machine is a stable mechanism that *generates a regularity*. An SNM will be a stable chance setup or mechanism that generates a probability (or distribution). The best examples of SNMs, unsurprisingly, are classical gambling devices: dice on craps tables, roulette wheels, fair coin tossers, etc. For these and many other kinds of chance setup, we can deduce their chancy behavior from their setup's structure and the Stochasticity Postulate. Not all genuine objective chances have to be derivable from the SP, however. We will consider examples of objective chances that are simply *there*, to be discerned, in the patterns of events.

Nevertheless, any objective chance should be thought of as tied to a well-defined chance *setup* (or reference class, as it is sometimes appropriate to say). The patterns in the mosaic that constitute Humean chances are regularities, and regularities of course link one sort of thing with another. In the case of chance, the linkage is between the well-defined chance setup and the possible outcomes for which there are objective probabilities.

Some linking of objective probabilities to a setup, or a reference class, is clearly needed. Just as a Humean about laws sees ordinary laws as, in the first instance, patterns or regularities—in the mosaic,

---

[1] The intuitive idea here is that of unpredictability, the non-existence of a discernible pattern in the events that permits a "gambling system" to make better forecasts than the objective chance itself does. For finite sets of events, it may not be possible to give a formal analysis of random-lookingness, and certainly I will not try. See (Earman, 1986, chapter 8). But I do want to note that my proposal is not that average people's judgments are definitive of random-lookingness. For large sequences and collections, "unusual" strings of events (e.g., 7 or more heads in a row when coin flipping) become more and more "probable" in the mathematical sense, but may nevertheless surprise non-statisticians and cause them to question the genuine randomness of the process.

whenever F, then G—so the Humean about chances sees them as patterns or regularities in the mosaic also, albeit patterns of a different and more complicated kind: whenever S, $P(A) = x$.

Two further comments on the notion of "chance setup" are needed. First, "well-defined" does not necessarily mean non-vague. "A fair coin is flipped decently well and allowed to land undisturbed" may be vague, but is nevertheless a well-defined chance setup in the sense that matters for us (it excludes lots of events quite clearly, and includes many others equally clearly). Second, my use of the term "chance setup," which is historically linked to views best thought of as propensity accounts of chance (e.g., Popper, Giere, Hacking), should not be taken as an indication that my goal is to offer a Humean theory that mimics the features of propensity theories as closely as possible. Rather, making chances conditional on the instantiation of a well-defined setup is necessary once we reject Lewis' time-indexed approach. For Lewis, a non-trivial time-indexed objective probability $Pr_t(A)$ is, in effect, the chance of A occurring given the instantiation of a big setup: the entire history of the world up to time t. Since I reject Lewis' picture of the world unfolding in time in accordance with chancy laws, I don't have his big implicit setup.[2] So I need to make my chances explicitly linked to the appropriate (typically small, local) setup.

Again, I do not insist that the chance-bearing "outcome" must come after the "setup" (though almost all chances we care about have this feature). Here is a possible example that goes the other way in time. Suppose that a historian has to try to transcribe a large amount of old, faded papyrus manuscripts. It occurs to her to train up a neural network-based program to do the job, saving human work-hours

---

[2] To clarify, what I reject here (or at least, wish not to presuppose) are two parts of this Lewisian picture: that the world *unfolds in time*, and that it does so governed by *chancy laws*. The former notion, which seems in tension with a B-series perspective on time (despite the fact that Lewis was definitely a committed B-theorist), seems to lend support to the view that any past event should be held (by us now, to the future of that event) to have objective chance = 1, which I reject. (Even if one accepts a growing block view of time, one may still reject the ascription of chance = 1 to all past events.) The latter notion I eschew because I do not want to commit myself to a Humean account of laws of nature. There may be universally applicable chance *rules* in our world—say, something like quantum mechanical chances—but I wish to not presuppose that it is correct to think of such rules as laws of nature.

without sacrificing accuracy. To train up the network she writes a lot of ancient Greek text on papyrus-like paper, then artificially ages/ smudges/damages the pages to mimic the way the real manuscripts are damaged. After some training up, the network is able to correctly transcribe an ancient Greek text on papyrus with high accuracy. At that point, she may announce: "The probability that a letter on the papyrus is a *rho*, given that my program scans and declares it to be a *rho*, is 0.94." Here the "outcome" (scanned letter is a *rho*, or is not a *rho*) clearly exists before the chance setup (papyrus scanning) associated with it. This probability may still be a perfectly good and useful Humean chance: what matters is whether the pattern of outcome-facts matches the chances in the appropriate way, not whether the outcome occurs before or after the "setup" conditions in time.

To understand the notion of patterns in the mosaic, an analogy from photography may be helpful. A black-and-white photo of a gray wall will be composed of a myriad of grains, each of which is either white or black. Each grain is like a particular "outcome" of a chance process. If the gray is fairly uniform, then it will be true that, if one takes any given patch of the photo, above a certain size, there will be a certain ratio of white to black grains (say 40%), and this will be true (within a certain tolerance) of every similar-sized patch you care to select. If you select patches of smaller size, there will be more fluctuation. In a given patch of only 12 grains, for example, you might find 8 white grains; in another, only 3; and so on. But if you take a non-specially-selected collection of 30 patches of 12 grains, there will again be close to 40% whites among the 360 total grains. The mosaic of grains in the photo is analogous to the mosaic of events in the real world that (partially) ground an objective chance such as, e.g., the chance drawing a spade in a well-shuffled deck.[3] In neither case does

---

[3] I say "partially ground" here because, arguably, the chances of card games played with well-shuffled decks are examples of SNMs. The effective randomization provided by human card-shuffling is what leads, via the deterministic unfolding of the game (given human decisions such as to draw or hold pat), to the reliable random-looking macroscopic pattern of draws of a spade. The chance is thus grounded in part on lower-level facts (about human physiology and the mechanics of card shuffling), not just on the pattern of card-draw outcomes. But the latter pattern does play a part: if (somehow, perhaps for no explicable reason) the frequency of drawing a spade from a shuffled deck

one have to postulate a propensity, or give any kind of explanation of exactly how each event (black, white; spade, non-spade) came to be, for the chance (the grayness) to be objective and real.

Of course, like photos, patterns in the mosaic of real world outcomes can be much more complex than this. There can be patterns more complex and interesting than mere uniform frequencies made from black and white grains (not to speak of colored grains). There may be repeated variations in shading, shapes, regularities in frequency of one sort of shape or shade following another (in a given direction), and so on. (Think of these as analogies for the various types of probability distributions found to be useful in the sciences, e.g., Poisson processes.)

There may be regularities that can only be discerned from a very far-back perspective on a photograph (e.g., a page of a high school yearbook containing row after row of photos of 18-year-olds, in alphabetical order—so that, in the large, there is a stable ratio of girl photos to boy photos on each page, say 25 girls to 23 boys). This regularity may be associated with an SNM—it depends on the details of the case—but in any case, the regularity about boys and girls on pages is objectively there, and makes it reasonable to bet "girl" if offered a wager at even odds on the sex of a person whose photo will be chosen at random on a randomly selected page.

Finally, we should think of the entire collection of chance-making patterns discernible in the events of the actual world—i.e., the HM— as determining a Best System of chances: a collection of propositions specifying chance setups and the probabilities of outcomes in them, where the overall collection has a maximal combination of simplicity, strength, and "fit."

With this clause, the specification of what objective chances *are* becomes complete, much as it did for Lewis. Up to this point we have implicitly characterized certain sorts of probability functions (linking them to distinct setups and to the patterns in the events falling under those setups). But what conceptually brings together all and only *the*

were consistently 1/5 instead of 1/4, over all human history, then the objective chance might not be 1/4 despite what the structure of the SNM seems to make highly probable.

objective chances is the Best System competition: "The chances are what the Best System says they are."

Since we should expect that chances are rather ubiquitous and to be found at many different ontic levels, we should not expect a gain in simplicity to outweigh a loss in strength nearly as much as seems plausible for Lewis' BSA account of laws of nature: the Best System of chances will be a very big collection of propositions indeed, though it will also undoubtedly contain things, like the QM-derived chances (as we will see in chapter 6), that are powerful and compactly speci-fiable. We may be tempted to call some of these propositions "chance laws," particularly those, like QM probabilities, that intuitively seem like candidates for inclusion in a Lewisian Best System of laws + chances. But there is nothing ontologically special, at bottom, about these chances compared to other kinds such as the chances of drawing a spade.[4] Over and above the intuitive grasp we have of simplicity, strength, and fit, what binds together the chances in a Humean Best System? Their aptness for guiding reasonable credences *via* PP, which we will demonstrate in chapter 4.

From now on, I will call this kind of chance that I am advocating "Humean objective chance" (or HOC for short).[5] But please keep in mind that the Humeanism only covers chance itself; not laws, causa-tion, minds, epistemology, or anything else.

3.2. Examples

Not every actual frequency, even in a clearly defined reference class, is an objective chance. Conversely, not every chance setup with a definite

---

[4] As we saw in chapter 1, if we try to understand "chancy laws" as primitives, concep-tually disconnected from actual events, it becomes hard or impossible to say what the content of (a statement of) such a law is. Forming part of a Best System remedies this problem, at the price of undercutting the idea—which I think we never truly grasped anyway—of chance laws "governing the unfolding of events."
[5] I will indiscriminately use "HOC" to refer to some particular objective chance, to the overall phenomena of Humean chances, or to the Humean *theory* of chance being elaborated and defended in this book. Context will always make clear which use is intended.

HOC need correspond to a large reference class with frequencies matching the chances. I will illustrate the main features of Humean objective chances through a few examples, and then extract the salient general features.

3.2.1. Chance of 00 on a Roulette Wheel

I begin with an example of a classic gambling device, to illustrate several key aspects of HOC. The objective chance of 00 is, naturally, 1/[the number of slots], and this is also the chance of each of the other possible outcomes. What considerations lead to this conclusion? (We will assume, here and throughout unless otherwise specified, that the future events (and past events outside our knowledge) in our world are roughly what we would expect based on past experience.) First of all, presumably there is the actual frequency, very close to x. But that is just one factor, arguably not the most important. (There has perhaps never been a roulette wheel with 43 slots; but we believe that if we made one, the chance of 00 would be $\frac{1}{43}$.)

Consider the type of chance setup a roulette wheel exemplifies. First we have spatial symmetry, each slot on the wheel having the same shape and size as every other. Second, we have (at least) four elements of *randomization* in the functioning of the wheel/toss: first, the spinning (together with facts about human perception and lack of concern) gives us randomness of the initial entry-point of the ball, i.e., the place where it first touches. The initial trajectory and velocity of the ball is also fairly random, within a spread of possibilities. The mechanism itself is a good approximation to a classical chaotic system—that is, it embodies sensitive dependence on initial conditions. Finally, the whole system is not isolated from external perturbations (gravitational, air currents, vibrations of the table from footfalls and bumps, etc.), and these perturbations also can be seen as a further randomizing factor. The dynamics of the roulette wheel and ball are fairly Newtonian, and it is therefore natural to expect that the results of spins with so many randomizing factors, both in the initial conditions and in the external influences, will be distributed stochastically but fairly uniformly over the possible outcomes (number of slots). And this expectation is amply confirmed by the actual outcome events, of course.

The alert reader may be concerned at this use of "randomness" and "randomizing," when these notions are surely bound up with the notion of chance itself (and maybe, worse, a propensity understanding of chance). But recall that for the Humean about chance, all randomness is product randomness.[6] Randomness of initial conditions is thus nothing more than stochastic-lookingness of the distribution of initial (and/or boundary) conditions, displaying a definite and stable distribution at the appropriate level of coarse-graining. The randomness adverted to earlier in my description of the roulette wheel is just this, a Humean-compatible aspect of the patterns of events at more-microscopic levels. Here we see the Stochasticity Postulate in action: it grounds our justified expectation that roulette wheels will be unpredictable and will generate appropriate statistics in the outcomes.

By assigning chance (1/[number of slots = N]) to each roulette wheel spin outcome, for any reasonable N, the Humean approach allows considerations of simplicity and strength to outweigh the consideration of matching actual frequencies. Suppose, contrary to fact, that 100-slot roulette wheel spins are very numerous in our world history, and the actual frequency of 00 outcomes in such wheels is not precisely .01, but rather .0099876625. This kind of small-scale frequency tolerance is characteristic of HOC. But the tolerance is by no means unlimited, as it is for propensity and hypothetical frequency accounts. If actual history contained many billions of 100-slot roulette wheel spins, and the frequency of 00 were .009850, while the frequency of single-0 were .010150, with all other slots having frequencies extremely close to .01, then presumably the Humean Best System would not specify all the objective probabilities as equal to .01; instead it would include a special-case clause for single-0 and 00 outcomes on 100-slot wheels, equating the OC with the actual frequencies (which are already simple and need no "smoothing off").

---

[6] The term "product" randomness is perhaps unfortunate, since it seems to imply that the randomness involved has been produced by some "random process." HOC rejects any such assumption.

3.2.2. Good Coin Flips

Not every flip of a coin is an instantiation of the kind of stochastic nomological machine we implicitly assume is responsible for the fair 50/50 odds of getting heads or tails when we flip coins for certain purposes. Young children's flips often turn the coin only one time; flips where the coin lands on a grooved floor frequently fail to yield either heads or tails; and so on. Yet there is a wide range of circumstances that do instantiate the SNM of a fair coin flip, and we might characterize the machine roughly as follows:

1. The coin is given a substantial upward impulse, so that it travels at least a foot upward and at least a foot downward before being caught or bouncing;
2. The coin rotates while in the air, at a decent rate and a decent number of times;
3. The coin is a reasonable approximation to a perfect disc, with reasonably uniform density and uniform magnetic properties (if any);
4. The coin is either caught by someone not trying to achieve any particular outcome, or is allowed to bounce and come to rest on a fairly flat surface without interference;
5. If multiple flips are undertaken, the initial impulses should be distributed randomly over a decent range of values so that both the height achieved and the rate of spin do not cluster tightly around any particular value. (One way to achieve this condition, of course, is to let ordinary untrained humans do the flipping.)

Two points about this SNM deserve brief comment. First, this characterization is obviously vague. That is not a defect. If you try to characterize what is an *automobile*, you will generate a description with similar vagueness at many points. This does not mean that there are no automobiles in reality. Second, here too the "randomness" adverted to is meant only as random-lookingness, and implies nothing about the *processes* at work. For example, we might instantiate our SNM with a very tightly calibrated flipping machine that chooses (a) the size of the initial impulse, and (b) the distance and angle off-center of

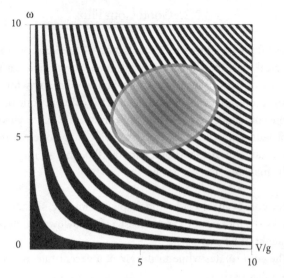

Figure 3.1. Coin flip initial condition space with flip outcomes shaded white or black. In the oval region, in which all reasonable initial spin and upward velocities fall, shading indicates density of the distribution of initial conditions.

Source: Diaconis, P. (1998).

the impulse, by selecting the values from a pseudo-random-number-generating algorithm. In "the wild," of course, the reliability of nicely randomly-distributed initial conditions for coin flips is, again, an aspect of the Stochasticity Postulate.[7]

Figure 3.1 adapted from (Diaconis, 1998), illustrates this, on a Newtonian-physics model of coin flipping. Initial conditions with ω (angular velocity) and V/g (vertical velocity) falling in a black area land heads; those in white areas land tails. (The coins are all flipped starting heads-up.) From the SP we expect the initial angular velocities

---

[7] (Sober, 2010) discusses a coin-flipping setup of the sort described here, following earlier analyses by Keller and Diaconis based on Newtonian physics. Sober comes to the same conclusion: if the distribution of initial conditions is appropriately random-looking (and in particular, distributed approximately equally between ICs leading to heads and ICs leading to tails), then the overall system is one with an objective chance of 0.5 for heads.

and vertical velocities to be scattered in a random-looking distribution over some central region of the square (not an *even* distribution, but rather random-looking in the sense of not having any correlation with the black and white bands). When this is the case, the frequency of heads (black bands) and tails (white) will be approximately 50/50.

Box 3.1 The *Method of Arbitrary Functions* Approach

Both the roulette wheel and the coin flipper are instances of dynamical systems that satisfy the requirements for Poincaré's "method of arbitrary functions," and a good number of philosophers have in recent years made this method the foundation of their analyses of objective probabilities of one sort or another (e.g., Von Plato, 1983; Myrvold, 2012; Strevens, 2011). A brief discussion of these approaches here may be helpful, to see how they relate to HOC and to my notion of stochastic nomological machines.

The beauty of the method of arbitrary functions (MAF) approach, to its proponents, is that it allows probabilities that are *objective* in some sense (and the precise sense varies among the proponents) to be applied in the context of physical systems that are taken to be deterministic. The roulette wheel and the coin-flipping systems we have discussed earlier fit perfectly into the MAF paradigm.

Where the term "arbitrary functions" fits into the picture is in the presumption that there is some probability distribution over the possible initial conditions of the system—which may be taken as an epistemic probability function (Myrvold), or as a mere representation of facts about actual initial conditions frequencies (Strevens). The upshot either way is what we saw for the coin-flipper dynamics earlier: almost any arbitrary distribution of initial conditions, if not too bumpy or lumpy in a way correlated with the outcomes, will yield the familiar outcome probabilities for the system. Strevens and Myrvold both use the example of a two-color wheel of fortune. Figure 3.2, from Strevens, illustrates two such initial distribution functions. The system is simpler than Diaconis' coin-flip

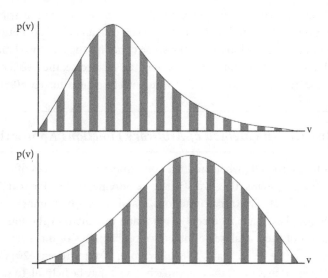

Figure 3.2. Different spin speed distributions induce the same probability for *red* outcome.
Source: Adapted from Strevens (2011), p. 347.

model in having only one initial condition, velocity, determining the outcome.

The MAF approach to probability from deterministic systems clearly has much in common with my discussion of SNMs. The Stochasticity Postulate that I advocate can be seen as claiming that it is a widespread and reliable fact about our world that the distribution of *actual* initial conditions for SNMs—if such a thing exists; or the distribution*s* of such initial conditions, if they exist but vary over time/place—belong to the "non-funny-looking" class of functions that entail the customary outcome probabilities.

Myrvold's development of the MAF idea, however, stays firmly in the realm of credence. Starting from initial conditions distributions understood as credences, Myrvold's account generates only what he calls "epistemic chances," which he distinguishes from true objective chances. Although they have a strong degree of objectivity in the sense that any agent with *reasonable* priors will turn out to assign roughly the same epistemic chances to outcomes, they are not tied to actual initial condition facts in the way that HOCs (or Strevens' initial conditions frequency distributions) are. So in a world with the same dynamics as ours, but with quite odd initial conditions for coin flips or wheel of fortune spins, while HOC entails that the objective chances of outcomes for these gambling device setups are different, Myrvold's epistemic chances remain the same as they are in actuality.[a]

---

[a] Strevens' account also entails, I believe, that the chances in such a world would be different, although possibly not in precisely the same way as the HOCs would differ (since his account has no Best Systems component). Myrvold's epistemic chances would not differ. But over time good Bayesian agents in such a world might come to have "weird" credences over possible initial conditions, and thus end up having credences that are in tune with actual frequencies anyway. The agents would learn, in effect, not to apply PP to epistemic chances—which arguably makes the "chance" part of "epistemic chance" slightly misleading.

3.2.3. The Biased Coin Flipper

The coin flip SNM just described adds little to the roulette wheel case, other than a healthy dose of vagueness (due to the wide variety of coin flippers in the world). But the remarks about a coin-flipping machine point us toward the following, more interesting SNM. Suppose we take the tightly calibrated coin flipper (and "fair" coin) and: make sure that the coins land on a very flat and smooth, but very mushy surface (so that they never, or almost never, bounce); try various inputs for the initial impulses until we find one that regularly has the coin landing heads when started heads-up, as long as nothing disturbs the machine; and finally, shield the machine from outside disturbances. Such a machine can no doubt be built (probably has been built, I would guess), and with enough engineering sweat can be made to yield as close to chance = 1.0 of heads as we wish.

This is just as good an SNM as the ordinary coin flipper, if perhaps harder to achieve in practice. Both yield a regularity, namely a determinate objective probability of the outcome heads. But it is interesting to note the differences in the kinds of "shielding" required in the two cases. In the first, what we need is shielding from conditions that bias the results (intentional or not). Conditions 1, 2, 4, and 5 are all, in part at least, shielding conditions. But in the biased coin flipper, the shielding we need is of the more prosaic sort that many of our finely tuned and sensitive machines need: protection from bumps, wind, vibration, etc. Yet, unless we are aiming at a chance of heads of precisely 1.0, we cannot shield out these micro-stochastic influences completely! This machine makes use of the micro-stochasticity of events, but a more delicate and refined use. We can confidently predict that the machine would be harder to make and keep stable, than an ordinary 50/50-generating machine. There would be a tendency of the frequencies to slide toward 1.0 (if the shielding works too well), or back toward 0.5 (if it lets in too much from outside).

3.2.4. The Radium Atom Decay

Nothing much needs to be said here, as current scientific theory says that this is not an SNM at all: it has no deterministic underlying dynamics that takes initial conditions and turns them into outcome events. In this respect it is very unlike gambling systems, and some will wish for some explanation of the reliable decay frequency patterns we find (see Box 1.5 in chapter 1). Whether we can have one or not remains to be seen.[8] Other philosophers will want to try to reduce *all* objective chances to this sort. Whether they can have their way will be the subject of section 2 of chapter 5.

Although the radium atom's half-life may not be associable with a deterministic, structured SNM, it is nevertheless associated with a strong/simple pattern in actual events, namely the pattern encoded in the combination of Schrödinger's equation and the "Born rule" for deriving probabilities for measurement outcomes. Together these two

---

[8] Bohmians think they already have it—and their explanation turns the radium atom back into a deterministic SNM in my sense, and invokes the SP with respect to particle initial position distributions (see C. Hoefer, 2011).

equations provide an extremely strong and simple systematization of vast numbers events. This being the case, the frequency tolerance we should grant to quantum-level events like radium decays is stronger than that of N-slotted roulette wheels. A large enough mismatch between actual frequency and candidate $1/N$ chances, displayed over a very large number of actual trials, we said, might force us to include a special clause for the outcomes displaying the mismatch. But the incredible simplicity and strength of QM-derived chances point toward another possibility: if there is a small mismatch between, for example, the theoretical half-life of radium and the pattern actually instantiated in decay events, this may not be enough to warrant a special clause for radium's half-life; *even if the number of decay events instantiating the actual pattern is infinite.* The increase in "fit" would not be worth the loss in simplicity and strength. But this frequency tolerance, too, is limited: if the difference between the actual half-life and the QM-derived half life is significant enough, a special clause or law for radium decay *will* make it into the Best System. When we deduce the validity of PP for HOC in section 4.1 of chapter 4, the delicate balance of frequency tolerance and intolerance, characteristic of Humean chance, will prove very important.

Many Humean objective chances—especially the paradigm cases—will be associable with an SNM whose structure we can lay out more or less clearly. But we should expect that many other Humean chances will not have such a structure. If they exist, out there in the wild, they exist because of the existence of the appropriate sort of pattern in actual events. But the class of "appropriate" patterns will not be rigorously definable. There will be no clear-cut, non-arbitrary line that we can draw, to divide genuine objective probabilities on one side, from mere frequencies on the other. Our last example illustrates such chances-in-the-wild.

3.2.5. The 9:37 Train

Let's assume that there is no SNM that produces a chance regularity (if there is one) in the arrival time of my morning train. Is there nevertheless an objective chance of the 9:37 train arriving within +/– 3 minutes of scheduled time? Perhaps—it depends on what the pattern/distribution of arrival times looks like. Is it nicely random-looking while overall fitting (say) a nice Gaussian distribution, over many months? Is the distribution stable over time, or if it shifts, is there a nice way to capture how it slowly changes (say, over several years)? If so, the implied chance rule may well make the cut for inclusion in the Best System for our world. On the other hand, suppose that the pattern of arrivals failed to be random-looking in two significant ways: it depends on day of the week (almost always late on Friday, almost always on time on Monday, etc.), and (aside from the previous two generalizations), the pattern fails pretty badly to be stable across time, by whatever measure of stability we find appropriate. In this case, it probably does not make sense to say there is an objective chance of the train being on time—even though, taking all the arrivals in world history together, we can of course come up with an overall frequency.

When we discuss the deduction of the Principal Principle (PP), we will see why such mere frequencies do not deserve to be called objective chances. Stability is the crucial notion, even if it is somewhat vague. When there is a pattern of stable frequencies among outcomes in clearly defined setups (which will often be called *reference classes*), then guiding one's expectations by the Humean chances (which will either be close to, or identical with, the frequencies) will be a strategy demonstrably better than the relevant alternatives.

3.3. The Best System of Chances

I think Lewis was right to suppose that a Humean approach to objective chance should involve the notion of a Best System of chances—though not a Best System of laws + chances together. Now it is time to say more about this idea.

Those who favor the BSA account of laws are welcome to keep it; my approach to chance does not require rejecting it. All I ask, as noted earlier in chapter 2, is that we allow that, in addition to whatever chances the BSA laws may provide, we can recognize other Humean chances as well, without insisting that they be part of (or follow from) the chancy laws. Then, in addition to a Best System of laws, there will also be a Humean Best System of chances, which I will now characterize.

Lewis was able to offer what appeared to be a fairly clear characterization of his Best Systems, with his criteria of strength, simplicity, and fit. By contrast, my characterization of chance best systems may appear less tidy from the outset. But there is a good justification for the untidiness. First, the "best" in Best System means best *for us* (and for creatures relevantly similar). The system covers the sorts of events we can observe and catalogue, and uses the full panoply of natural (and artificial) kind terms that we use in science and in daily life. Pattern regularities about coins and trains may be found in the Best System, not only regularities about quarks and leptons. Since we are not trying to vindicate fundamentalism or reductionism with our account of chance, but rather to make sense of real-world uses of the concept, there is no reason for us to follow Lewis in hypothesizing a privileged physical natural kinds vocabulary.

In any case, closer inspection of Lewis' theory destroys the initial impression of tidiness. Simplicity and strength are meant to be timeless, objective notions unrelated to our species or our scientific history. But one suspects that if BSA advocates aim to have their account mesh with scientific practice, these notions will have to be rather pragmatically defined.[9] Moreover, simplicity and strength are simply not clearly characterized by Lewis or his followers. We don't know whether initial conditions, giving the state of the world at a time, should count as one proposition or as infinitely many (nor how to weigh the reduction of simplicity, whatever answer we give); we don't know whether deterministic laws are automatically as strong as can be, or whether instead some added chance-laws may increase strength at an acceptable price; and as Elga (2004) has noted, the notion of "fit" certainly cannot be the one Lewis proposed, in worlds with infinite numbers of chance events.

I propose to retain the three criteria of simplicity, strength, and fit—understanding fit along the lines sketched by Elga, which we will

---

[9] This is argued by Van Fraassen in *Laws and Symmetry* (1989), chapter 1.

look at in the following—but now applied to systems of chances alone, not laws + chances. These three notions can be grasped for chance systems alone at least as clearly as they can be grasped for laws + chances systems. For Lewis, *strength* was characterized in terms of the amount of the overall Humean mosaic "captured" by a system. Although never spelled out, this notion of strength appears to be unduly aimed at the capturing of petty details (e.g., the *precise* shape, mass, and constitution of every grain of sand on every beach, etc.).[10] When considering chance systems, the capturing of quite particular detail is neither desirable nor achievable, and strength is instead most naturally understood in terms of how many different *types* of phenomena the system covers. Strength should be determined by the net domain of the system's objective probability functions. So if system 1's domain includes everything from system 2's domain, plus more, then 1 beats 2 on strength. Where two systems' domains fail to overlap, it may be difficult to decide which is stronger, since that may require adjudicating which builds strength more: covering apples, or covering oranges (so to speak). But fortunately, since our systems' chances are not constrained to have the kind of simplicity that scientists and philosophers tend to hope that the true fundamental laws have, this difficulty is easily overcome: a third system that takes (e.g.) system 1 and adds the chances-about-oranges found in system 2 will beat both in terms of strength.

What about simplicity? The value of simplicity is to be understood not in terms of extreme brevity or compact expression, but rather in terms of (a) elegant unification (where possible), and (b) user-friendliness for beings such as ourselves. But elegance is not such an overriding virtue that we should consider it as trumping even a modest increase in strength bought at the expense of increased untidiness. In fact, I tend to see the value of (a) as really derivative from (b): user-friendliness. User-friendliness is a combination of two factors: *utility* for epistemically and ability-limited agents such as ourselves, and

---

[10] Maudlin has criticized Lewisian approaches to laws for this apparent emphasis on the trivial (personal conversation).

confirmability (which, ceteris paribus, elegant unification tends to boost).[11]

Finally, I need to say a few words about "fit." Lewis' idea of fit aimed at including *everything* that happens in the world, and assigning one global fit-level to each candidate Best System that has any chances at all, namely: the chance that the system assigns to the full course of actual history. Elga (2004) has shown convincingly that this idea is unworkable if there is an infinite number of chance events (according to a candidate system) in the actual world, which of course there may well be if either space or time is infinite.[12] The problem is simple: if there is any such infinity of candidate-chancy events in our world, then the Lewis fit of any system that assigns them chances is *zero*. Any two systems assigning them chances, no matter how different, will share this precise fit of zero, given Lewis' definition. And Elga shows that the problems cannot be overcome simply by adopting a non-standard analysis and assigning infinitesimal fits.

Further problems can be seen in Lewis' definition. For example, there is no recipe for how to multiply up the chance events if the world lacks (as we think it does) an absolute time order. For another, it is highly prejudiced against the inclusion of chances for chance setups that occur an infinite number of times: Any system according to which only a finite number of events are chancy in our world will win out, on fit, over any system that assigns chances to an infinite number of events. It seems clear that a completely new approach to "fit" is required.

Elga proposes that the Humean not try to measure fit against the entire set of chancy events, but instead look at certain "test" propositions, namely the true, "simple" propositions (relative to the candidate Best System). These are, essentially, propositions specifying the outcomes of actual trials (single trials, or a finite set of specific trials—where

---

[11] Here I am assuming a standard Bayesian notion of confirmation, based on the kind of prior probabilities about chances widely shared (in an implicit sense) among scientists.

[12] Infinite sequences of chance events in time, as we saw in section 1.2 of Chapter 1, may be problematic if our world is as we believe it to be, but even so, our world *may* instantiate never-ending sequences of chance events (perhaps, for example, radium decays)—just not infinite sequences of *coin flips*.

by "trial" we mean a medium-long sequence or medium-sized set of instantiations of the chance setup). If one system assigns higher probability to an appropriately weighted majority of these test propositions than does the other, then it has better fit.

Elga's proposal allows us to compare fitness scores for rival candidate systems that cover exactly the same chance setup *types*, even if an infinite number of events occur within the domain of one or more chance rules. This resolves the zero-fit problem, but it does not yet give us a general definition of fit that can be used to compare systems with non-identical sets of chance setup types.

The way to handle this, I think, is to recognize that "fit" is a notion that comes into play, in the Best System competition, only at the stage where considerations of simplicity and strength have already narrowed down the field of competitors to a set of systems that all agree on what types of chance setups should be covered.[13] Once that stage is reached, we will have a set of competitor systems that all cover the same events, but which, by specifying different *specific* chance rules to cover these events, earn different fitness and simplicity scores.

At this point, simplicity and fit will trade off in at least two different ways. First, there may be systems that gain simplicity at a potential cost of reduced fit. An example would be a system that covers all gambling devices with appropriate physical and spatial symmetries, having N possible outcomes, with a uniform chance rule that the chance of each possible outcome is $1/N$. Suppose that (say) coin flips are a type of event where, in actuality, a rule that *Pr(Heads) = 0.51* is much closer to the actual frequency than the $1/N$-implied chance of 0.5. Then the uniform $1/N$ system is simpler than one that breaks out a special rule for coin flips (keeping the $1/N$ chance rule intact for all other symmetric gambling devices), but its fit is a bit worse.

---

[13] A *caveat* here: the phenomena potentially covered by a rule must display enough statistical regularity and systematizability to be plausibly covered by a chance rule in the first place, and this requirement is of course an aspect of *fit*. In this sense, *minimal fit* is a first requirement for any candidate chance rule to be even entered into a candidate system. But once this criterion is met, then further impact of the differences in fit between different candidate chance rules enters only after the competition has already narrowed down to competing systems that all cover the same types of chance setups.

A second way in which simplicity and fit trade off is between competitors that have exactly the same chance rule types. The simplicity of a system that assigns chance 0.5 to both heads and tails, in coin flips, is arguably higher than one that assigns chance 0.51 to heads and 0.49 to tails; and certainly higher than one that assigns chance 0.50966745 to heads and 0.49033255 to tails.[14] But the fit of the latter two systems, for test propositions concerning coin flips, may be higher, and especially so for the third mentioned system.

Given these ways in which simplicity and fit trade off against each other, what determines the victor? Here we must acknowledge some facts about Best System accounts (whether of chance or laws or both) that have been the source of frequent objections from opponents of the view. It is unclear that the notions of simplicity and strength can be plausibly held to be fully objective (as opposed to being subjective and/or anthropomorphic notions). Even if they were, it is a further and equally dubious thing to claim that the correct way to weight the trade-offs between them is fully objective. Rather than attempt the quixotic task of defending the full objectivity of the parameters of the Best System competition, I prefer to minimize the impact of the subjective elements by noting two points, and then accept, as almost certainly harmless, whatever ambiguity remains.

The first point is that, since my aim is not to offer an account of laws of nature, but rather only of objective chances, there is no pressure to make it seem plausible that the Best System competition will spit out a system arguably similar to what science has already delivered to us. In the case of laws, defenders of the Lewisian program are keen to make plausible that the Best System competition ends up delivering something like a single all-encompassing physical theory, like the fabled Theory of Everything that physicists desire; or if not a single theory, at least a fairly small collection of fundamental and wide-ranging laws. But since I divorce HOC from any particular theory of laws, and depart from a stance of skepticism about objective chances (if understood in a primitivist or propensity-theory sense), I feel no

---

[14] The reason why the system with $Pr(Heads) = 0.50966745$ is certainly less simple has to do with a pragmatic aspect of simplicity that we will discuss later: calculational simplicity for the user of the chances.

pressure at all to engineer the Best System competition so as to yield a small number of chance rules. So simplicity, understood in terms of having a simple and compact expression, is much less important to me than strength. I would expect the Best System for our world to be a rather large book, with many many chance rules covering domains of many kinds. (We will see positive arguments why this should be so in the following, and in the course of the next two chapters). If the "finalists" in the competition are all systems with a very large number of chance rules, covering by and large all the same phenomena, then if the imperfect objectivity of strength, simplicity, fit, and the trade-offs between them leave us with no clear victor among these finalists, it won't really matter. We could take any one of them and consider its rules to be giving us genuine Humean objective chances for our world.

The second point is that the trade-off between fit and simplicity is strongly constrained by the overarching goal of my approach to understanding objective chance: the correct theory of chance has to be such that the objective chances it delivers to us are *demonstrably* apt for guiding credence in the way captured by the Principal Principle. Given this aim, how much can fit be sacrificed for increased simplicity? Not much at all. "Chances" that mislead us concerning what to expect, either often, or seldom but by a large amount, will clearly not be apt guides for credence; so any way of cashing out our three notions (simplicity, strength, and fit) that could lead to such a system being declared victorious is a non-starter.

The right way to understand my pragmatic Humean Best System approach to chance will be clarified and refined further in what follows and in the remaining chapters, but I hope that the picture is starting to become clear. There is no particular reason why the book of rules given by the Best System for our world should be short; but neither is there reason to think that it could or should be an endless compendium of trivial frequency facts. It is meant to be "Best" for *us*—that is, for epistemically limited creatures embedded in world history. So the chances should be both discoverable by us (when circumstances are right) and *usable* by us—a point that will emerge clearly in chapters 5 and 6. The notions of simplicity, strength, fit, and their trade-off values are not fully objective, and that presumably means there can be no single clear victorious Best System for our world, even in the eyes of a Laplace's demon with complete information about our Humean Mosaic. But the "finalist" systems, among which it is impossible to find a single Best one, will have to be mostly overlapping, in both the types of phenomena they cover and the chances they ascribe to them. And that is all the objectivity that we need, or could rightly demand.

3.4. More about the Contents of the Best System

Objective chances are a "guide to life," and one that ideally we can get our hands on by observation, induction, and experimentation. Lewis tried to distance himself from agent-centered values in describing his criteria of simplicity and strength because he wanted his account to mimic, as closely as possible, the physicist's notion of fundamental laws. But for an account of chances alone, there is no need to insist on this kind of Platonic objectivity. There may be such deep laws in our world, and some of them may even entail probabilities. But there are also lots of other chances, dealing with mundane things like coins, cabbages, card games, and diseases.

HOC thus offers an empiricist account of objective chance, but one more in the mold of Mach than of Lewis. This may seem like a disadvantage, since the anti-metaphysical positivism of Mach is almost universally rejected, and rightly so. But here I endorse none of Mach's philosophy, not even his view of scientific laws as economical summaries of experience. If we are convinced that propensity theories and hypothetical frequentist theories of objective probability are inadequate, as we should be, then how can objective probabilities be salvaged at all? HOC offers a way to do so, but its objective chances may appear to be more agent-centric and less Platonically objective than those postulated by the proponents of other theories. (In the next chapter we will see that the agent-centric aspects of HOC are virtues, when it comes to justifying the PP for my Humean chances.)

For a Humean about chance, what chances exist, how much of the overall mosaic they "cover," and how well they admit systematization are all questions whose answers depend on the contingent specifics of the universe's history. And while we have come to know (we think) a lot about that history, there is still much that we have yet to learn. Despite our relative ignorance, there are some aspects of a Best System for our world that can be described with some confidence.

Earlier we discussed roulette wheels and I mentioned that for any well-made wheel with N slots (within a certain range of natural numbers N), each slot's number has a probability of $1/N$ of winning on each spin. This is an example of the kind of higher-level chance fact that we should expect to be captured by the Best System for our world. It goes well beyond frequentism, since it applies to roulette wheels with few or zero actual trials, and it "smooths off" or "rounds up/down" the actual frequencies to make them fall into line with the symmetries of the wheels. But still, this chance regularity is just a regularity about roulette wheels. We can speculate as to whether the Best System for our world is able to capture this regularity as an instance of a still-higher-level regularity: a regularity about symmetrical devices that amplify small differences in initial conditions and/or external influences to produce (given the SP) a reliable symmetric and random-looking distribution of outcomes over long sequences of trials. Given what we know about the reliability of certain kinds of mechanisms, and the reliability of the stochasticity of the input/boundary conditions for many such mechanisms, this seems like a solid speculation. I would not want to try to articulate a full definition of such SNMs, which have as sub-classes roulette wheels, craps tables, lottery ball drums, and so forth. But we do not have to be able to specify clearly all of the domains of objective chance, in order to have confidence in the existence of some of them.

The full domain of chance includes more than just gambling devices, however, even at the macro-level. There may or may not be an objective chance of the 9:37 train being on time, but there probably is (thanks to the biological processes of sexual reproduction) an objective chance of a given human couple having a blue-eyed child if they have a baby, and there may well be an objective chance of developing breast cancer (in the course of a year) for adult women of a given ethnicity aged 39 in the United States. I say "may well," because it is not automatically clear that in specifying the reference class in this way, I have indeed described a proper chance setup that has the requisite stability of distribution, syncrhonically and diachronically. The problem is well known: if there is a significant causal factor left out of this description, that varies significantly over time or place, or an irrelevant factor left in, then the required stability may not be found in the actual patterns of events.[15] If the required stability is present, though, then there is a perfectly good objective chance here, associated with the "setup" described.

It may not be the *only* good objective chance in the neighborhood, however. Perhaps there are different, but equally good and stable statistics for the onset of breast cancer among women aged 39 *who have had children and have breast-fed them for at least 6 months*. There is a tendency among philosophers to suppose that if this objective chance exists, then it cancels out the first one, rendering it non-objective at best. But this is a mistake. The first probability is perfectly objective, and correct to use in circumstances where one needs to make predictions about breast cancer rates and *either* (a) one does not know about the existence of the second objective probability, or (b) one has no information concerning child-bearing and breast-feeding for the relevant group. There *is* a sense in which the second probability can "dominate" over the first, however, if neither (a) nor (b) is the case.

PP, with admissibility correctly understood, shows us this. Suppose we are concerned to set our credence in

A: Mrs. K, a randomly selected woman from the New Brunswick area aged 39, will develop breast cancer within a year

and we know these three objective chances:

$$X1: \Pr(B.cancer|woman\ 39,...) = x_1$$
$$X2: \Pr(B.cancer|woman\ 39\ \&\ has\ breast\text{-}fed,...) = x_2, \text{ and}$$
$$X3: \Pr(B.cancer|woman\ 39\ \&\ has\ not\ breast\text{-}fed,...) = x_3$$

---

[15] As I argue in "Humean Effective Strategies" (2005), the requisite stability may also fail just due to statistical "bad luck"—one should not think that presence/absence of causal factors will explain everything about the actual statistical patterns.

and for the population, we have all the facts about which women have had children and have breast-fed them. With all of this packed into our evidence E, we *cannot* use PP in this way:

$$Cr(A|X1 \& E) = Pr(A) = x_1$$

Why not? Because since our evidence E contains $X2$, $X3$, *and* the facts about which women have breast-fed children (including Mrs. K), our evidence contains information relevant to the truth of A, which is not information that comes by way of A's objective chance *in the X1 setup* (the one whose invocation we are considering). So, this information is all jointly inadmissible. By contrast, we *can* apply PP using $X2$, because all our evidence is admissible with respect to that more refined chance. Knowing $X2$, we also know that $X1$ is not relevant for the truth of A for cases where we know whether a woman has breast-fed or not.[16]

What I have done here should remind us of the advice that empiricist/frequentists have given for ages, to set the (relevant) objective probability equal to the frequencies *in the smallest homogeneous reference class for which there are "good" statistics*. But we have, hopefully, a clearer understanding of what "homogeneous" means here, in terms of the chance setups that make it into the Best System's domain; we see that we may be able to apply the objective chance even if the relevant events form a reference class too small to have good statistics—namely, if the Humean chance is underwritten by a higher-level pattern that gives coverage to the setup we are considering; and finally, we see why this advice does not automatically undercut the claim to objectivity of probabilities for larger, less-homogeneous reference classes.[17]

In order for the more-specific and less-specific chance to exist side by side, it must be the case that there are no dramatic fluctuations over

---

[16] That a person belongs to the X2 reference class entails that the person belongs to the X1 reference class, hence the latter must be admissible evidence relative to X2. (Thanks to Allan Gibbard for pointing this out).

[17] It might seem that the Best System aspect of our Humean account of chance should rule out such overlapping chances: the system should choose the *best* of the two competing chances. In this case, that would perhaps mean jettisoning the chance *X1* and retaining the *X2* and *X3* chances. In some cases that may be correct, but not in general. Considerations of discoverability and utility (applicability) will often be enough (together with the existence and stability of the right sort of outcomes in history) to demand the retention of more-general chances alongside more-specific ones, in the Best System.

time and/or space, in the reference class of all 39-year-old women of
X1, in the proportions of women who breast-feed.[18] If there were, then
the patterns of breast cancer among the *X1* women would fail to have
the stability we require for an OC, and only the *X2* chances would remain. So if both exist, we can infer that there is a third objective prob-
ability: *Pr(Breastfeeds|woman-in-X1-ref-class)*. This may seem odd,
since breast-feeding is a *decision* freely made by women, not something "governed by chance." But this just goes to show, again, that we
should stay away from thinking of Humean chances as *governing*.

When considering the chances of unpleasant outcomes like
cancer, of course, we would typically like something even better than
one of these objective probabilities for setups with many people in
their domain. We would like to know *our own, personal* chance of
a certain type of cancer, *starting now*. The problem is that we can't
have what we want. There is good reason to doubt that this objective chance actually exists, in the best system for our world that an
omniscient being would lay out. Since a given person's history does
not suffice to ground a chance-making pattern for cancer, for such
a chance to exist it would have to be grounded at a different level,
perhaps by reduction to micro-level probabilities. But even if this
reductionist objective chance exists as a consequence of the Best
System, we are never going to be able to know its value (not being
omniscient). So for the purposes of science, of social policy, and of
personal planning, such individualized objective chances may as
well not exist; and a philosophical account of chance that hopes to
be relevant to the uses of probability in these areas of human life can
forget about them.

Nonetheless, as we will argue in detail in chapter 7, it is surely true
that *part* of the Best System chances in our world are the chances
given to us by quantum mechanics.[19] It is a matter of controversy

---

[18] There is no clear-cut line to divide cases where some putative *X1* chances do have
the requisite stability, from cases where they do not. Again, the boundary between
clearly existing Humean objective chances and mere statistics is a vague one, like the
boundary between automobiles and non-automobiles.

[19] Remember: *I* am entitled to say this, because it does not hinge on whether determinism is true or false, or on whether there is ultimately some better, more-
fundamental theory for phenomena in our world. Philosophers who follow Lewis in
either (a) denying the compatibility of chance and determinism, or (b) analyzing chance
+ laws together in a big Best Systems package, are not entitled to this claim.

whether all phenomena whatsoever are covered by quantum mechanics (including its field theory extensions). But suppose this were true. Would the quantum chances then give us objective probabilities for every specifiable event, including each individual's personal chance of contracting each type of cancer in the course of the next year? In chapter 6 we will explore this question, and more generally how micro-level and macro-level chances should relate to one another.

Box 3.2 Does PP Really Need an Admissibility Clause?

Some writers on chance and credence claim that PP, rightly understood, needs no admissibility clause after all (see, e.g., Meacham, 2010).

How could that be? And are they right? My answer to the latter question is "No," although the idea has obvious attractions for those concerned by the vagueness of the admissibility clause as I formulate it.

The general idea for how PP might *not* need an admissibility clause can be easily seen by considering the elements of PP2 as Lewis understood them. Recall:

$$C(A|H_{tw}T_w) = x = Pr_t(A) \qquad \text{(PP2)}$$

The theory of chance T_w is said to be a giant collection of history-to-chance conditionals, essentially a function that takes a history of the world up to time t as input and outputs an objective chance distribution defined over some domain of propositions. (As I noted in chapter 2, Lewis and those working in his tradition often assumed that, by default, the domain of the chance theory is essentially unrestricted; more on this later.) And since its input, H_{tw} is a *complete* description of the world up to time t, T_w has at its disposal all the facts that any agent could have at time t (and much more than any real agent could have). Moreover, because of the way the Best System prescription works, T_w should be thought of as having "taken into account" the *entire* history of the world, future as well. So T_w is an expert credence guide that knows everything the agent may know, and more, and has taken into account the patterns of future events as well as past ones; there is just no way that an agent could have *better* information than what $H_{tw}T_w$ provides. Therefore, there's no need for an admissibility proviso for PP.

What about the standard examples used to motivate an admissibility clause, such as time travelers from the future or magic crystal balls, that one takes as providing trustworthy evidence concerning future chancy events? Well, all facts about them—what they reveal, whether their revelations are or are not correlated with events to their future—are part of H_w, the full history for w, which was the supervenience base for the Best System competition; that competition has already taken into account all such facts! And since *your* seeing an image in a crystal ball, or having conversed with an apparent time traveler, is in your past, it is included in H_{tw}, which is input into T_w in order to grind out the objective chance $P_t(A)$. System knows best, so you don't need to worry that you might somehow have a better epistemic perspective than does the theory of chance T_w. Admissibility is just not a concern.

One problem with this argument is that it ignores the role that simplicity can play in the Best System competition. Consider the chance rule that governs coin flips, and let's suppose that there are only a few instances in world history where a time traveler tells a person what the result of a future coin flip will be. Then in the full world history, there may be a 100% correlation between time-traveler-predicted flip outcomes and the actual outcomes. This pushes in favor of establishing a special chance rule for such flips, or perhaps adding a clause to the ordinary coin flip chance rule exempting flips that get time-traveler-predicted. But such modification of the system reduces either simplicity and/or usability, with zero gain in strength and at most infinitesmal improvement in fit. Maybe the Best System will swallow the loss of simplicity, and let the system's rules "take into account" the pronouncements of time travelers, as the no-admissibility-needed position seems to assume. But then again, maybe it won't. And if it does not, then application of PP2 will say that for an ideal rational agent, $C(A|H_{tw}T_w) = Pr_t(A) = \frac{1}{2}$. But this is just wrong.

As it happens, given the way I have stressed that (in HOC) simplicity (in the sense of keeping the *number* of chance rules low) counts for little compared to strength and fit, arguably, this sort of problem is unlikely to affect the Best System. (It is, however, a serious problem for the Lewis way of understanding Best Systems.) But in HOC, a different feature may make jettisoning the admissibility clause problematic in a different way. That feature is the (possible/probable) *patchy* nature of the coverage of the rules in the Best System of our world. We can see the problem easily if we consider a case discussed by Donald Gillies, the chances of a motorcycle accident happening to his niece Francesca who was a teenager in Rome.

Let's suppose that there is an objective chance of teenage moto drivers in Rome having an accident (in the space of a year's time), and that chance is 0.08. Francesca and her parents know this, but she argues that they need not worry too much because *her* chance of having an accident is much lower; she is a much more sensible and cautious person than the average Roman teenager. Now, when it comes to HOC as I lay it out, it is quite possible that the Best System does contain a chance rule for teen moto accidents in Rome, but does *not* contain a rule that covers only Francesca. Is she irrationally violating PP when she sets her credence in having an accident far lower than 0.08? I would say *no*, and the reason is precisely that her epistemic situation is one of having inadmissible information for the application of that chance rule. She knows that she is more cautious and sensible than most of her peers, whose behaviors and perchances jointly underlie the chance rule in question. This meets the test for inadmissible information as I defined it in chapter 2.

Situations like Francesca's are probably fairly ubiquitous, under HOC; and indeed in real life, when we do have some indication of what the objective chances for certain things are (usually, by knowing some frequency data), we nevertheless tweak our credences up or down depending on other factors that we take to be relevant. So an admissibility clause for PP is necessary after all.

3.5. Summing Up

Chances are constituted by the existence of patterns in the mosaic of events in the world. These patterns are such as to make the adoption of credences identical to the chances rational in the absence of better information, if one needs to make guesses or bets concerning the outcomes of chance set-ups (as I will show in chapter 4). Some stable, macrocscopic chances that supervene on the overall pattern are explicable as regularities guaranteed by the structure of the assumed chance setup, together with our world's micro-stochastic-lookingness (SP). Not all genuine objective chances have to be derivable from the SP, however. The right sort of stability and randomness of outcome-distribution, over all history, for a well-defined chance setup, is usually enough to constitute an objective chance. Moreover, setups with few actual outcomes, but the right sort of similarities to other setups having clear objective chances (e.g., symmetries, similar dynamics, etc.) can be ascribed objective chances also: these chances supervene on aspects of the Humean Mosaic at a higher level of abstraction. The full set of objective chances in our world is thus a sort of Best System of many kinds of chances, at various levels of scale and with varying kinds of support in the Humean base. What unifies all the chances is their ability to play the role of guiding credence, as codified in the PP.

Box 3.3   A Cosmic Dilemma for HOC?

The following objection has been put to me by Aidan Lyon in correspondence, and I have encountered similar objections made by other philosophers over the years. The core of the objection can be thought of as a dilemma: If the supervenience base of the Best System of chances is the *whole* HM of our cosmos, then trivial things like cancer rates and train arrivals are too small a part of the Mosaic for it to make sense to compromise the simplicity of the system by including them; but on the other hand, if only the part of the HM around here is taken into account as the supervenience base, then if there are (by chance) any statistical flukes in, say, the decay rates of certain rare elements in our part of the HM, our system will wrongly declare the objective chances of decay to be in accord with the statistical flukes (which mistake would not be made if we stuck to using the whole HM as supervenience base).

The latter problem is perhaps avoidable if we interpret 'around here' to mean "in the observable universe," i.e., the chunk of the HM that lies in the backward light cone of the end of human history (or the backward light cone *now* if you prefer a growing block). Then the decay rates in that large swath of spacetime can't be seen as a mere statistical fluke; if they differ non-trivially from rates in other parts of the full HM, then there deserves to be a special rule for such decays in our region of spacetime in the overall Best System anyway. But if we take this still-rather-large swath of spacetime as the relevant supervenience base for the Humean chances, it looks like the problem of the first horn reappears: since our planet is such a small speck in that swath, and all human history such a tiny temporal stretch on the scale of billions of years, surely the Best System should prize simplicity over the miniscule extra strength to be had by adding rules for cancers and gambling games and so forth, and eschew any chance rules for these things?

The pragmatic, for-creatures-like-us nature of HOC, as I analyze it, is the key to resolving this apparent dilemma. Since we are not analyzing the chances as part of a package deal including the laws of nature, the intuitive pull that a Lewisian may feel, to give simplicity real importance and keep the total number of laws (including chance-laws) low, does not exist for my analysis. As I noted in this chapter, the Best System competition puts strength—defined in terms of *number of* types *of chance setup covered in the system*—and fit ahead of simplicity, although not absolutely so in all regards. Strength *for beings like us* is improved by including more and more chance rules—as long as the rules concern well-defined setups that are actually instantiated a large number of times and display clear-cut statistical regularities. (In order for *fit* to be respected as a criterion, rules should not be added for setups which are such that the actual events falling under them do not display any sufficiently stable statistical regularities.) And to give simplicity *for beings like us* its due weight, very complicated, hard to use, or hard to discover chance rules will be omitted from the Best System unless they cover, with excellent fit, such widespread phenomena that to omit them from the system would incur too great a loss of strength.

When the Best System competition is understood along these lines, the cosmic dilemma is avoided. The system's chance rules will surely cover many terrestrial phenomena, despite what a small speck our world is in the cosmos. The risk exists that a special chance rule for Raritanium decay rates may get into the system, grounded on the decay rates on Earth, even though the terrestrial decay rates look like a statistical fluke compared with the patterns elsewhere in the cosmos; but this is a positive feature, not a bug. When we see how the validity of the PP can be demonstrated for HOC, in chapter 4, the advantages of HOC's way of balancing strength, simplicity, and fit will become even clearer.

4

Deducing the Principal Principle

In this chapter I will show how, if objective chances are as the HOC view specifies, it follows that PP—or something extremely close to it—is a constraint on rational credence. Lewis claimed: "I think I see, dimly but well enough, how knowledge of frequencies and symmetries and best systems could constrain rational credence" (D. Lewis, 1994, p. 484). Michael Strevens and Ned Hall have claimed that Lewis was mistaken, as either there is *no way at all* to justify PP, on any view of chance (Strevens, 1999), or no way for a Humean to do the job (Hall, 2004). I will try to prove these authors mistaken by direct example, offering a few comments along the way on how they went astray.

In fact, I will try to establish the validity of PP, or something extremely close to it, twice over: first with a "consequentialist" argument, and then with an "a priori" argument. I will give a compact statement of each argument here, with the full development and defense of the arguments occupying most of the rest of this chapter. But first, a word of warning and a disclaimer. The word of warning: the arguments in this chapter will be rather more complex and multifaceted than anything found in previous chapters. Readers who feel, like Lewis, that they already see dimly, but well enough, how Humean chances do indeed constrain rational credence, should feel free to skip this chapter (although some elements of this chapter play a role in some of the arguments of chapter 5). The disclaimer: what the arguments of this chapter will demonstrate is that it would be irrational for an agent with no inadmissible information to *not* have $Cr(A|XE) \approx x$. That is, the arguments will show that it is irrational to have credences, conditional on XE, that differ *non-trivially* from the objective chances. The wiggle room left open is, I will argue, pragmatically unimportant, but arguably useful (as we will see in chapter 5).

The consequentialist argument for PP can be summarized like this: The way that chances supervene on the HM in HOC guarantees that for chance setups S that occur a large number of times, the frequency of A-outcomes will be close to the chance of A, and that agents with no inadmissible information will have no grounds to expect any particular sort of deviation (+ or −) of the frequency from the chance, at any time or place. Therefore agents know that betting strategies based on the chances should lead to winning, or at least approximately breaking even. (Here, I assume that one can choose which side of the bet to take, and betting against an opponent who may accept odds different from those corresponding to the objective chances.) Therefore, the agents know that their credence in A outcomes should be x, or very near to x.

The a priori argument for PP is roughly this: to have credence y in A, a possible chance outcome of a setup S, is to implicitly commit oneself to there being no advantage to either side, in an indefinite series of bets on whether A in S, at odds of $(y:(1 - y))$. But given the tight connection HOC imposes between frequencies and chances when the number of instantiations of a setup S grows arbitrarily large, the agent knows that if y is not approximately equal to x, betting consistently on one side at these odds will lead to a certain loss (or certain gain), which means they are not fair odds after all. Thus, given the way we understand credences, the agent has contradictory beliefs if her credence in A is significantly different from x. Rationality requires eschewing inconsistent beliefs, so if the agent has a credence (conditional on the chance being x) at all, it should be equal to x, as PP demands.

4.1. Deducing the Reasonableness of PP

The key to demonstrating the validity of PP for Humean chances rests on the fact that the account is a "sophistication" of actual frequentism. For the purposes of this chapter, we can think of HOC as modifying simple actual frequentism by:

A. Requiring that outcomes not only have an actual frequency (or limit, if infinity is contemplated), but also that the distribution of outcomes "look chancy" in the appropriate way—stability of distribution over time and space, no great deviations from the distribution in medium-sized, naturally selected subsets of events, etc. This is something a frequentist should insist on, in any case.[1]

B. Allowing higher-level and lower-level regularities (patterns), symmetries, etc., to "extend" objective chance to cover setups with few, or even zero, actual instances in the world's history (often with the help of the SP and the notion of an SNM).

C. Anchoring the notion of chance to our epistemic needs and capabilities through the Best Systems aspect of the account.

D. Insisting that the proper domain of application of objective chances is intrinsically limited, as we will see in detail in chapters 5 and 6.

With these ideas in mind, we can begin the deductions. First, we'll see how it is possible to give an argument along what Strevens (1999) calls "consequentialist" lines, to show that adjusting one's credences to the chances is an optimal strategy (assuming that one's beliefs about the Humean chances are correct). Then we will show that someone who understands what Humean objective chance is, and takes herself to know some HOCs, yet does not conform to PP, is being *incoherent*. This is just what Howson and Urbach tried to show for von Mises chances (see chapter 1, section 1.3). It turns out that a close cousin of the Howson and Urbach argument can be given for Humean chances; and it avoids the problem that their argument suffered due to its appeal to infinite limits. The frequency *intolerance* of HOC proves crucial to both demonstrations of the rationality of PP.

4.1.1. The Consequentialist Argument

First, let's recall the Howson and Urbach way of defining subjective degrees of belief, or credences (chapter 1, section 1.3): one has degree

---

[1] Von Mises (1981) insisted on something of this nature, but in order to make it mathematically tractable in the way he desired, he had to make the unfortunate leap to hypothetical infinite "collectives."

of belief p in outcome A if one believes that there would be no advantage to either side of a bet on A at odds $p:1-p$.[2] To conform to PP is to make one's wagers on matters that fall inside the domain of objective chance by setting the credence equal to the chance (as always, when one's other knowledge is all admissible). PP can thus be shown pragmatically rational by showing that, when one's utility depends on making accurate guesses concerning the outcomes of chance setups, in the absence of any better information one does best to guess using the objective chances as one's guide. Of course, PP only advises you to set your credence equal to what you *believe to be* the objective chance. The pragmatic argument tries to establish that, if your belief about objective chance is *correct*, then following PP maximizes your expected utility in a prediction/betting situation such as that we are considering. This is then enough to demonstrate the rationality of PP as a cognitive strategy, given the usual understanding of practical rationality.[3]

PP advises us to set our level of credence in an outcome type A, under setup conditions S, equal to [what we take to be] the objective chance x of A, as long as we have no better (inadmissible) information to go by. What we need to demonstrate is that there is an objective sense in which the recommended level of credence x is better than any other that we might adopt instead. Here our assumption will be that we are dealing with a simple, time-constant objective chance of A in the setup S. Not all objective chances in the Best System need be

---

[2] Or rather: if one would so believe, if one were made aware of the meaning of the relevant terms ("bet," "odds"). Clearly, a person can have a degree of belief without having been introduced to the world of gambling. See (Eriksson & Hájek, 2007) for discussion of some of the pitfalls of using betting-related dispositions to define credences. While they make many good points against traditional definitions of degrees of belief, I think that it remains true that there is a tight connection between subjective degree of belief and what an agent does or would consider "fair betting odds" concerning a proposition's truth. This connection is captured better by Howson and Urbach's definition than it is by the earlier proposals criticized by Eriksson and Hájek. See Box 1.1 in chapter 1 for further discussion.

[3] Rather than spell out a particular theory of practical rationality, I think it is best to stick here to our shared implicit understanding of the kind of rationality at issue: the kind of rationality that tells us (*ceteris paribus*) to prefer X to Y if our beliefs indicate that X will probably bring us greater utility (be it money, happiness, or whatever) than Y. Every specific, rigorous fleshing out of practical rationality can be challenged with paradoxes, counterintuitive consequences, and so forth; but those issues do not affect the use I will make here of the core notion.

like this, of course, but our argument will carry over to less-simple chance laws and distributions more or less directly and obviously. Let's suppose that this is a typical objectively chancy phenomenon, in the sense that S occurs very many times throughout history, and A as well. Then our Humean account of chance entails that the frequency of A in S will be approximately equal to x, and also that the distribution of A-outcomes throughout all cases of S will be fairly uniform over time (stable), and stochastic-looking. If the first (frequency) condition did not hold, x could not be the objective chance coughed up by the Best System for our world—world history would then undermine a chance-value of x. If the stability condition did not hold, then either our Best System would not say that there exists an objective chance of A in S, or it would say that the chance is a certain function of time and/or place—contrary to our assumption.[4] If the stochastic-lookingness condition did not hold, then again either the Best System would not include a chance for A in S, or it would include a more complicated, possibly time-variable or non-Markovian chance law—contrary to our assumption.[5]

Therefore, we know the following: at *most* places and times in world history, if one has to guess at the next n outcomes of the S setup, then if n is reasonably large (but still "short run" compared to the entire set of actual S instances), the proportion of A outcomes in those n "trials" will be close to x. And sometimes the proportion will be greater than x, sometimes less than x; there will be no discernible pattern to the distribution of greater-than-x results (over n trials) versus less-than-x results; and if this guessing is repeated at many different times, the

---

[4] Actually, the stability condition could fail for A in S, if the chances in S are derived from a higher-level chance rule and S is just an anomalous sub-setup within this larger rule. However, the same considerations we're applying here work at the higher level to entail that such anomalous sub-setups are the exception rather than the rule, and hence can be set aside. My goal in this chapter is not to prove that one *always* does best when using the OCs, only to show that there is no reason to think that one can do better by doing something different.

[5] For example, it might be that after any two consecutive flips turn up heads, the frequency of heads on the next flip of the same coin is 0.25, on average, throughout history. If this were so, then (depending on how other aspects of the pattern look) we might have a very different, non-Markovian chance law for coin flips; or a special sub-law just for cases where two flips have come up heads, and so forth.

average error (sum of the deviations of the proportions of A in each set of n trials from x) will almost always be close to zero. [6] Notice that we have made use of the ordinary language quantifiers "most" and "almost always" here, and we have *not* said anything about getting things right "with high probability" or "with certainty, in the limit," or anything of that nature.

This already makes it intuitively clear that betting at odds determined by PP will most of the time lead to approximately breaking even, rather than suffering a near-certain financial loss (as happens, for example, if one bets on heads at odds of 3:2 in favor). Or to put the utility of PP in a more positive light, imagine that one bets with an opponent who proposes stakes $j{:}k$ (you put up j, he puts up k) for a series of bets on a chance process where the outcome A has objective chance p. You get to choose whether to bet on A or $\neg A$. The strategy PP recommends is to bet on A if $p > j/(j+k)$, on $\neg A$ if $p < j/(j+k)$, and on either side (or neither—refuse to bet) if they are equal.

This is the setup of Strevens' (1999) discussion of PP as a "winning strategy" for short- or long-run "games" ("games" being sequences of bets on the same side, at constant odds). If the opponent offers stakes far enough away from the odds dictated by the objective chances, then intuitively one is guaranteed to win in most games, by following the PP-derived betting strategy. In his Appendix B, Strevens offers a proof that that following PP is a "winning strategy" in the [medium-] short run if the relative frequency f of A is close to the objective chance x, where "close" means that f is closer to p than to $j/(j+k)$. Our pragmatic argument shows that "most" games of the Strevens type will be won by the user of PP, at most places/times in history, whenever the game has a medium-sized length and the difference between $j/(j+k)$ and p is significant.[7]

Since Strevens maintains that no justification of PP is possible at all, under any theory, what does he make of the short-run argument

---

[6] The minimum size of the short-run of n trials needed for the argument here to work depends in part on the nature of S and on the size of x. For example, if $x = 0.5$, $n = 100$ gives a very decent-sized run in which the frequency of A will usually be somewhat close to x (e.g., between 40 and 60 per 100); if $x = 0.003$, then n must be much larger.

[7] As in the previous footnote, the notion of a "significant" is partly a function of the length of the sequence of trials in the game.

for PP in his Appendix B? He views it as a case of "close, but no cigar" because he thinks that there can be no *guarantee* that the antecedent condition—that the frequencies of A in most games be close to the objective chance—will be satisfied, in a chancy world (pp. 258–259). (Like Hall, and most other authors, he thinks of objective chances as being, as propensity theorists maintain, compatible with any results whatsoever over a finite set of cases.) Strevens does consider the possibility that satisfying these conditions might be built into the definition of chance directly in one way or another (as, in effect, is done in Humean chance), but each of the specific ways Strevens considers is different from HOC. For example, he considers a definition of chance that stipulates that *most* short-run games over all history yield frequencies close to the chances. This definition leaves open the possibility of patches of history where the frequencies differ substantially from the chances in most games, so Strevens asks: How can we know we do not live in such a bad patch? HOC, however, does rule out such bad luck for us: if a region of spacetime such as that covered by all human history has frequencies of A quite different from those in the rest of world history, then HOC will say that there is a special chance law for our region of spacetime (or a clause exempting our region from having any chance of A, if the problem in our region is wild instability of frequencies). The pragmatic best-system nature of HOC, with its built-in variety of frequency intolerance, saves the day.

A second possibility Strevens considers is a long-run frequency definition, with a clause stipulating that there are *no* "bad patches" anywhere. Against such a theory Strevens rightly objects that it gains nothing because it makes it impossible to demonstrate that the chances exist. But again, despite a certain resemblance, HOC is quite different from Strevens' target here. To be sure that there exist Humean objective chances, and to know their values, limited observations combined with ordinary inductive practices suffice. (More on the epistemology of HOC and its use of induction later, in section 4.2). Despite a valiant attempt to rule out all possible non-trivial justifications of PP for theories of objective chance, Strevens in effect fails to consider the logical subspace of theories of chance occupied by our Humean best system account.

♥

Let's get back to the pragmatic argument itself. What we have seen so far shows that, if one has a practical need to guess the proportion of A-outcomes in a non-trivial number of instances of S, then guessing x is not a bad move. Is there a better strategy out there, perhaps? Well, yes: it is even better to set one's credences equal to the actual frequency of A *in each specific set of* n *trials*—that way, one never goes wrong at all. This is basically the same as having a crystal ball, and we already knew that guiding one's credences by a reliable crystal ball is better than any other strategy. We can associate this idea with a competitor "theory" of objective chance, which we will call "Next-n Frequentism" (NNF). NNF-chances are clearly very apt indeed for playing the PP role. To the extent that they diverge from the Humean chances in value, someone who needs to predict the next n outcomes would be well advised to set their credences equal to the NNF chances. Unfortunately, unlike Humean chances, which can be discovered inductively (see section 4.2), you really would need a crystal ball to come to know the NNF-chances. Since no reliable crystal balls seem to exist, and there is no other way to arrive at this superior knowledge, we can set aside NNF-chances as irrelevant.

Is there any other *fixed* proportion x' that would be an even better guess than x? As long as the difference between x' and x is non-trivial, the answer to this question is, clearly, "No." Any such x' will be such that it diverges from the actual proportion of A in a set of n trials in a certain direction (+ or −) more often, overall, than x does. And its average absolute error, over a decent number of sets of n trials, will *almost always* be greater than the average absolute error for guessing x.

"Almost always" is not "always." It is true that a person might set her credences to x', guess accordingly, and do better in a bunch of sets of n trials than an x-guesser does. PP can't be expected to eliminate the possibility of bad luck. But the fact that *most of the time* the x-guesser accumulates less error than the x'-guesser is already a good start on our project of establishing the reasonableness of PP.

So far we have established that setting one's credences via PP is a better strategy, in our hypothetical game of guessing proportions of A-outcomes, than is setting credence equal to some *other* fixed level $x' \neq x$. In section 4.1.4 we will consider how this rationality carries over (to the extent it does) to single guessing events, to chance setups with few actual instances, and so forth. But at this point I have to acknowledge a limitation of this pragmatic argument: it shows the PP-derived strategy to be better than any rival *fixed x'* strategy, but does not thereby demonstrate that PP beats every other *possible* strategy for setting credences or guiding bets. As Marc Lange has pointed out, in the hypothetical game we are discussing, a person might set his or her credences for each set of *n* trials "randomly," or capriciously by pure whim, and nevertheless do better than the PP-user—even in an arbitrarily large sequence of bets. Nothing hangs on the word "whim" here; the hypothetical agent could set credence by rolling an *n*-sided polyhedron, by consulting a random number table, and so on.[8] While we recognize that success of such a strategy is possible, we still intuitively feel that the person who sets her credences by such a strategy, *even though she understands what chance is, and believes she knows what the objective chances are,* is irrational. Can nothing further be said to demonstrate the irrationality of such a non-PP strategy?

One is immediately tempted, of course, to point out that while such PP-beating success is *possible*, it is *highly unlikely*. What can we mean by this "unlikely"? Given the nature of HOC, for some variants of the "arbitrary" strategy class, it may be possible to argue that there is in fact a (Humean) objective chance of beating PP using such a strategy; it will of course be very low. That is good, but not yet enough. What we want is to be able to conclude that we should have *low subjective credence* in our being successful if we use one of these arbitary strategies, because we want PP-following to have the highest *expected utility* of any strategy. And to get this low subjective probability out of the low objective chance, we need to invoke PP. In the context of an overall argument aiming to establish the rationality of PP, this would amount to a vicious circularity. And in any case, for a strategy based on true *whim*, it will not be possible to argue that the HOC of success is low—because that chance will, most likely, simply not exist in our world.

---

[8] Lange and his students at UNC Chapel Hill presented this objection to me in their Kennan Summa seminar series in November 2006. I am grateful to Lange and all the participants for an extremely probing and useful discussion of HOC during the seminar.

Fortunately, answering Lange's challenge does not require *proving* that PP can be expected to have higher utility than a caprice-governed strategy. Rather, it is enough to point out that one can have no reason to expect, ahead of time, that such a strategy *will* do better, while one has some reason to expect that it will not. Conceptually, there is no linkage between setting credences by caprice and success (as, we have seen, there *is* in the case of setting credences equal to Humean chances using PP). On the other hand, the simple facts about how close frequencies in sets of n trials are to the chances, most of the time, give us reason to expect that caprice-governed credences will very often diverge from the actual frequencies. The counsel of practical rationality still clearly favors PP, even though the logical possibility exists of adopting credences (and hence betting odds) by caprice and emerging successful. The pragmatic justification of PP for HOC requires neither ruling out bad luck for the PP-follower, nor excluding good luck for the less rational!

Having accepted all this, it may still feel as though there really ought to be *something* further to say to demonstrate that those who eschew PP in favor of some other strategy, even though they (believe they) know the objective chances, are acting irrationally. And there *is* a further argument that can be made, to show precisely this. The Howson and Urbach argument for PP can be adapted to HOC. And we will see that, adapted to fit HOC, the argument goes through without requiring us to have beliefs about physically impossible infinite sequences of events.

4.1.2. The *A Priori* Argument: Howson and Urbach Redux

Recall, from chapter 1, the gist of Howson and Urbach's (1993) justification of PP for von Mises hypothetical frequency chances. To have subjective credence k in A is to believe that there is no advantage to either side of a bet on whether A occurs, at odds of $k{:}(1-k)$. Suppose that A is specified only as being an outcome of a [repeatable] chance setup, having objective chance x. Then to have credence k in A is implicitly to commit oneself to there being no advantage to betting on either side at these odds no matter how many times the bet is made, on successive trials of the setup. In particular, even in the limit as the number of bets goes to infinity, one believes these are the fair odds. But given how von Mises probabilities are defined, if one knows that the von Mises chance of A is x, then one knows that in the limit of infinite trials, the frequency of A in S will be exactly x. A standing bet on A, at anything other than odds of $x{:}(1-x)$, is bound to favor one side over the other in the wager. Therefore, one is incoherent—*literally, logically* incoherent—if one has credence $k \neq x$.

In chapter 1 I took this argument to task for two reasons. The first related to the hypothetical infinity of trials: in general, I argued, it is simply not possible to have beliefs about what *would* result if trials in a chance setup were repeated infinitely, because such repetition is not physically possible in a world like ours. The second related to the unlimited frequency tolerance of von Mises chances: the initial segment of a von Mises collective that corresponds to all trials that actually happen in human history may be anything whatsoever, without that affecting the "true" objective probability. Like robust metaphysical propensities, von Mises chances have unlimited frequency tolerance in any finite portion of a collective, so they cannot guarantee *pragmatic* success from use of PP.[9] But now, Humean objective chance never invokes infinite limits, and it certainly does not have unlimited frequency tolerance. So perhaps something very much like the Howson and Urbach argument can be used to show that the non-PP-conforming agent under HOC is *also* incoherent—literally incoherent.

Recall, again, that a subjective credence corresponds to a ratio of stakes at which one believes bets are fair. When the proposition at issue is one

[9] If Howson and Urbach's *a priori* argument for the rationality of PP were acceptable, then they could lay claim to the use of PP in order to mount a probabilistic argument for the pragmatic success of using PP. That is (and this is familiar from the usual dialectics on objective chance), they could point out that a sequence of events needed to produce a long initial segment whose frequencies differ significantly from the true chance is a kind of sequence that has a very low probability of occurring (*very* low). Therefore (by PP) it seems we should be extremely confident that such a thing will not occur, and therefore confident that using PP will indeed be a pragmatically successful strategy. But again, we *should* have these convictions *on pain of incoherence*, not because the expected success proveably *will* occur.

that falls under a chance setup, to have credence x in A is, in Howson and Urbach's words, "implicitly to commit onseself to there being no advantage" to either side in a series of bets on the chance setup having outcome A, at odds $x{:}(1{-}x)$, regardless of how short or long the series of bets turns out to be. Now, to know that S is a setup with a Humean objective chance x of producing outcome A does not entail knowing anything about how many times the S setup is run, in world history. (If one has independent means of knowing that, it may or may not be inadmissible evidence—set this aside for the moment). But it does entail knowing that there are strict limits to the frequency tolerance of the chance outcomes, and in particular, it entails knowing that as the number of actual trials increases, the closeness of the actual frequency to x must increase too (aside from possible temporary swings away). It is not possible for the actual frequency (AF) to stay resolutely distant from the HOC for S as the number of trials increases indefinitely, nor even for a very substantially long time (still much shorter than all of S's history), since that would constitute an *underminer* for that chance, entailing that the true best system assigns a different chance to A in S, or a time-variable chance that would restore closeness of the HOC to the AFs.[10,11]

In other words, to believe that the HOC of A in S is x is to believe that the actual frequency of A-outcomes in an indefinitely extended (though still finite) series of trials *cannot* be significantly different from x. But now we're home free: it is literally incoherent to have subjective credence x' in outcome A, if (a) one has no inadmissible evidence, i.e., your *only* rational grounds for taking yourself to know anything relevant to whether or not A will come to pass is your belief that the HOC of A is x; and (b) x' is more than trivially different from x. This is so because—just as in the Howson and Urbach argument—one is committing oneself on the one hand to bets at odds given by x' having no long-run advantage

---

[10] We will discuss undermining events, and what to say about them, in chapter 5.

[11] A possible objection here might be that, given the definition of a Best System, it might be possible for the AFs to differ from the HOC in some type of setup, even over an indefinitely long period of time, if the *simplicity* of the resulting system more than compensates for the lack of fit. The reply is that simplicity just does not have enough "weight" to make this happen. Worsening the simplicity of a candidate system by adding one new chance rule is a small imperfection, whereas indefinitely long *and significant* frequency-chance mismatch is a large imperfection.

or disadvantage, and on the other to the actual frequency of A in an indefinitely extended long run of trials being extremely close to x, which entails that a similarly long series of bets on one side of the issue at stakes $x'/(1-x')$ is sure to lose money. Contradiction.

So we are able to show, not only that using PP is a pragmatically rational strategy (the conclusion of 4.1.1), but also that it is logically incoherent *not* to apply PP, if one understands the nature of HOC. As with the pragmatic argument, the incoherence argument has perhaps more "slack" than a logician might like. Howson and Urbach could, as it were, drive x' arbitrarily close to x by going to the limit of infinity, and appealing to the properties of a von Mises collective. But HOC long runs, even if they are infinite somehow,[12] are not von Mises collectives. They cannot have arbitrarily long initial sequences that are non-random-looking or that deviate in a big way from the right frequencies, on the one hand (which lets us have the pragmatic justification of PP); and on the other, they *can* deviate *slightly* in long-run frequency from the HOC, even if infinite trials are actual. So in a strict sense my deduction of PP's rationality shows—in both parts—that it is foolish to set one's credences differently from what one takes to be the OC, where "different" means "non-trivially different."

How big is the "slack"? Not big enough to matter. This means something extremely small in quantum physics, but something (perhaps) several orders of magnitude larger in the context of card games. In whatever domain, the slack—defined by how much the actual frequencies might *in principle* differ from the HOCs, in an indefinitely large series of trials—is inherently restricted by the partly pragmatically defined nature of HOC. Recall, HOCs are aspects of the patterns in events *apt for guiding credence as indicated in PP*. Too-significant deviations are not permitted because they would undermine HOCs aptness for guiding credences (along *either* the pragmatic or the a priori line of justification sketched in the preceding). We see here a kind of circularity, but it is of the virtuous kind.

---

[12] In section 1.2 (of chapter 1) I did not claim that *no* chance setup can be imagined to run an infinite number of times in our world, just that certain important ones like coin flips cannot physically possibly do so. On the other hand, if something like quantum mechanics is true, then even after our world achieves "heat death," there will be events happening, e.g., electrons randomly forming from virtual-particle annihilation events or suchlike.

The demands of reason do not support PP to an arbitrary number of decimal places. But that is as it should be. Since the objective chances are *not*, in fact, hidden entities with numerically sharp values on the real interval [0,1], nor limiting frequencies in infinite collectives, arbitrary precision is not warranted by the nature of objective chance itself. The demands of reason *do* justify PP, however, with as much accuracy as can reasonably matter to us in any given domain.

Box 4.1 Sleeping Beauty: A Counterexample to PP?

The "sleeping beauty problem" is a delightful puzzle about rational credence that has been hotly debated over the past 20 years without a clear *consensus* resolution emerging so far. A recent paper by Nathan Moore (2017), which we will look at shortly, should change that situation.

Sleeping Beauty is the subject of a psychology experiment on credences. On Sunday night Beauty will be put into a deep coma-like sleep, and a fair coin will be flipped. If the coin lands heads, Beauty will be woken up on Monday and asked what her credence in H is. Then she will be told that it is Monday, and asked again what her credence in H is. She will then be put back to sleep for 48 hours and woken on Wednesday, and told that the experiment is over (and which way the coin landed—our psychologists are not cruel!). If the coin flip lands tails, then Beauty will be woken on Monday and asked questions, just as in the preceding. However, Beauty will then be given a sleeping drug that causes short-term amnesia: when next awoken, she will have no memory of having been woken on Monday morning or asked any questions. Beauty will then be woken on Tuesday morning also, and asked about her credence in H, then put back to sleep, to finally be awoken on Wednesday and (again) told what the coin flip outcome was.

Beauty is told all these relevant facts about the experiment, so she knows going in that any time she is awakened during the experiment, she will not know whether it is Monday or Tuesday. And finally, Beauty believes the coin flip will be fair and she respects the

PP, so on Sunday evening before she is first put to sleep her credence in H is $\frac{1}{2}$. The Sleeping Beauty puzzle question is this: when Beauty wakes up during the experiment, what **should** her credence in H be? If Beauty is the most rational Bayesian imaginable, how will she answer that first, Monday morning question?

There are two main camps in the debate about this puzzle: the halfers, who say that Beauty will answer "$\frac{1}{2}$," and the thirders, who say that she will answer "$\frac{1}{3}$." (There have been defenders of other answers, but we will not go into their reasonings.) Halfers have a simple and apparently cogent argument for their view: On waking up, Beauty knows only that she is in the experiment, but she knew in advance that she would be, and the coin flip result is independent of that fact. So, Beauty has no new information relevant to *Cr(H)*. This being so, she must still set her credence as PP apparently demands, to $\frac{1}{2}$. What's more, being told that it is Monday gives Beauty no new evidence relevant to H, since she knows that whichever way the coin landed, she was bound to be woken up on Monday. (The halfer who maintains that Beauty must answer "$\frac{1}{2}$" to both questions is called a "double halfer." Some halfers argue that Beauty should answer "$\frac{1}{2}$" before learning that it is Monday, but not afterward.)

Thirders have compelling arguments for their view, too. The first one that initially swayed me was this: If the experiment is run many times, Beauty will have many awakenings; in about $\frac{2}{3}$ of them, the coin will have landed tails (because tails causes her to be woken and interviewed twice). So Beauty can reasonably have credence $\frac{2}{3}$ that the coin has landed tails, since her situation is not different in any relevant way whether the experiment is run repeatedly or only once. Related arguments for the thirder answer can be devised, based on what bets Beauty would be rational to accept in the experimental setting.

The argument that has finally convinced me that thirders are right was first presented to me in a seminar paper by Nathan Moore, now contained in Moore (2017). The key to Moore's reasoning had already been independently discussed by P. Hawley (2013), though we did not discover that paper until much later (and Hawley, remarkably, sticks to defending the double halfer position!). The key is to consider Beauty's credence concerning whether or not it is Monday, when she first awakens (and knows that she is in the experimental situation, but—due to the amnesia protocol between Monday and Tuesday awakenings (when the latter is done)—not which day it is). What Moore demonstrates, from apparently unassailable premises, is that the double halfer answer commits one to saying that, on awakening, Beauty's credence that it is Monday must be 1. Beauty would have to be *certain* that it is Monday, despite knowing full well that it might well be Tuesday! This is clearly irrational, so the double halfer response is ruled out. Moore goes on to show that even if the halfer allows Beauty's credence in heads to change on learning that it is Monday, she will be committed to assigning very implausible credences in Monday to Beauty on her awakenings.

Readers are encouraged to read Moore's careful and clear exposition of how the halfer positions go wrong. What we need to consider here (and Moore also discusses) is this: does the thirder position's victory show that PP is somehow wrong after all?

If Beauty, on awakening, has credence $\frac{1}{3}$ in heads, then either she violates PP, or the admissibility clause of PP can be legitimately invoked in the SB scenario. Moore argues convincingly that the latter is the case. The key is again the linkage between Beauty's credence in Monday and her credence in heads. On Sunday evening Beauty had zero credence that it was Monday, and no inadmissible information regarding the coin flip, so she set her credence in heads to $\frac{1}{2}$. On awakening, however, Beauty must have an intermediate credence in Monday, somewhere between 0 and 1 (with, as Moore notes, the only intuitively reasonable credences in Monday being either $\frac{1}{2}$ or $\frac{1}{3}$). And a few lines of calculation using the probability axioms suffice to show that the value of Beauty's *Cr(Monday)* is *linked* to her credence in heads; the two cannot be set independently. Therefore, Beauty's knowledge that it *might be* Monday, and her specific credence level, are relevant to her credence in heads, and not by being relevant to the *objective chance* of heads. And, so described, we see that this is a clear-cut case of inadmissible "evidence." So we have, here, no counterexample to the rational validity of PP after all.

4.1.3. Strevens' Objection

Strevens (1999) gives a careful discussion of the Howson and Urbach argument for PP, and we need to consider his objection because it applies equally to my argument in the preceding. Strevens points out that a key premise of the argument is the one that takes us from an agent's having degree of belief k in the next trial of S having outcome A, to "implicitly committing oneself to there being no advantage" on either side at the $k:(1-k)$ odds for all the other trials of S, hypothetical or actual, needed to secure argument's conclusion. This premise, Strevens correctly notes, involves tacit appeal to something like a principle of indifference.

> Clearly, the "implicit commitment" is taken to be a consequence of our having "just the same information" about each and every toss in the sequences, namely the limiting frequency of heads in that sequence. Thus, though Howson and Urbach do not say so, they tacitly appeal to a principle of indifference along the lines of the one stated above. (1999, p. 262)

It is a recurrent theme in Strevens' article that many attempts to justify PP-like principles invoke one or another principle of indifference, and that these invocations can themselves only be justified by appeal to PP, or special cases of it. In the case of Howson and Urbach's argument, for example, Strevens notes that one could read them as appealing to a constrained, weaker principle of indifference, such as: "If all we know about a set of events is that they have been generated by the same physical setup, then we ought to assign the same subjective probability to each of these events being of some type E" (1999, p. 262). But when we consider what makes two instances of a chance process count as "generated by the same physical setup," we will soon realize that what we mean they have in common is precisely their objective chance of producing E. The principle invoked to support Howson and Urbach's argument would then become: Whenever two outcomes are produced by processes with the same objective chances, assign the same credences. And this is, Strevens says, just a special case of PP

(p. 263). If this is all that can be said, then it looks like the Howson and Urbach argument—and mine as well—are in trouble.

But I think the crucial premise invoked in our arguments is not, in fact, a version of the (objectionable, classical) Principle of Indifference, nor does one need to support it *via* an illicit appeal to a special case of PP. We can begin to see why this is so by noting that even the hidden premise formulated by Strevens in the preceding is not, in fact, a special case of PP. A special case of PP might be something like: "Whenever you know the objective chance of a particle decay is x, you should have subjective credence x in the decay's occurrence." But Strevens' principle does not tell you to set your credence equal to [what you take to be] the objective chance! It merely says: whenever two possible outcomes have the same objective chance, you should assign them the same subjective credence.[13] That leaves open the possibility that the chances could be 0.5, and your credences 0.75. Now, that would surely be foolish. And the Howson and Urbach argument, as well as my version of it, shows why. But that goes to show that the arguments, far from being circular, are in fact doing a very non-trivial job.

The implicit premise needed is by no means a special case of PP itself, but rather a much more basic principle of rationality, one that has (in fact) has nothing to do with chance setups per se. The principle formulated by Strevens, in his attempt to show how one might justify the "implicit commitment" part of the Howson/Urbach argument, did mention chance setups, and thus made it look like something unsettlingly circular must be going on. But I think that the true hidden premise should rather be formulated more like this:

True Principle of Indifference (TPI): *When one has no grounds at all on the basis of which to differentiate two propositions A and B, then if one has a subjective credence in the truth of A, one should assign the same credence to the truth of B.*

---

[13] Here, as usual, we are also assuming that one has no further/inadmissible information on which to base one's credences.

In our case, A and B will be propositions specifying the same outcome (e.g., heads) to distinct trials of the same chance setup. But there is nothing chance-related in the preceding True Principle of Indifference.

A couple of examples of TPI in action will help make this clear. Suppose you like David Lewis' papers; you think he's the greatest philosopher of the 20th century. In fact, in your experience, you agree with his conclusions most of the time. Now, I tell you I'm holding two different Lewis papers, that you've never read, behind my back. I call them A and B. I ask, what credence do you have in the proposition that A's conclusions are correct? You reply, "90%." I ask the same about B. Is it not clear that you are, in some way, irrational if you say, "Well, B is another ball of wax, fella! My credence in B's conclusions being right is just 50%." The irrationality does not *quite* get all the way to self-contradiction, but it is palpable nonetheless. And it is just this sort of irrationality that TPI rules out. This example involved intermediate credence levels, so it may still look suspiciously "chancy" (even though it should be clear that my holding two Lewis papers behind my back is not a genuine chance setup). So here is another example. Your partner tells you *A*: "There is a new jacket in the window at your favorite clothes shop." Let's suppose you believe this—as we usually do believe one another's assertions. Your partner goes on to say *B*: "Oh, and there is also a new fruit display at the supermarket." Let's suppose that you have not been out in a while, you know that both stores change their displays frequently, you trust your partner very much, can't think of any reason why s/he would lie about anything store-related, etc. In short: you have no information on the basis of which to differentiate your attitude toward claims *A* and *B*. Then you are irrational if you do, in fact, so differentiate. If you believe one, you should believe the other. If you assign intermediate credence to one, you should assign the same intermediate credence to the other. And finally—and most relevantly, for many other cases—if you simply do not assign any particular credence level to the one, you should also fail to assign a particular credence to the other.

TPI might be thought of as a Principle of Sufficient Reason for belief-differences: no difference in credence without some reason *for* the difference. Something like this is in fact a basic touchstone of human reasoning and argumentation. We often argue against each other by pointing out that the other is treating two cases differently (e.g., believing in X but not believing in Y) when the relevant evidence for both is identical. And we all recognize that this is a valid complaint, that such arbitrary asymmetries are to be avoided if possible. The a priori argument for PP does require some principle like TPI in order to go through, but it seems to me that the principle required is both plausible and very far from being, in any way, a special case of PP.

It is also, clearly, not the same as a Classical Principle of Indifference (CPI), which is invoked in order to turn ignorance into probability assignments. A CPI says something like: whenever the outcomes of a chance process can be divided into n outcome-types, and one has no information indicating that any of these n possibilities should be more likely than any other, then one ought to divide one's credence equally between them (i.e., set credence to $1/n$ for each). CPIs are rightly rejected nowadays, because it is recognized that there can be multiple ways of describing the outcomes of a chance process, and applying CPI using two different ways of categorizing outcomes can easily lead to contradictions (see Gillies 2000, chapter 3). But TPI is not like CPI, because instead of trying to turn ignorance into knowledge, it merely tells you to avoid turning ignorance into asymmetries (asymmetries that cannot be justified). In particular, TPI is perfectly satisfied if we withhold judgment entirely, and this is often the smart thing to do in situations of real ignorance.

So the Howson/Urbach argument and my version of it for HOC do not rely on any objectionable CPI, nor on any special case of PP itself. But considering Strevens' objections has brought to light another loophole in these arguments for PP. They cannot establish that *if* you know the objective chance of A, you *must* set your credence equal to it. You might instead opt to withold judgment—in operational terms, you may simply refuse to make bets or guesses, at any odds at all. Our arguments have not tried to show that this is irrational, and that is a virtue. But they do show that *if* one either must, or is willing to make guesses and bets concerning the outcomes of chance processes, and if one knows the objective chances (and nothing else relevant), *then* one is irrational if one does not adjust one's credences to match the objective chances. And that is to say that PP has been shown to be a principle of rationality.

4.1.4. Loopholes of the Arguments

Let's summarize our conclusions so far. PP tells you to set your credences equal to your best estimate of the objective chances. In general, the Humean objective chance is close to the actual frequency, and we have no way of predicting where and when any significant deviations may occur. Knowing what HOC is, we can be sure that the deviation between chance and frequency must become insignificant as the number of actual trials becomes indefinitely large. So: setting one's credences equal to the actual frequency is *pragmatically* better, at most times and places, than setting them equal to any other (significantly different) value; and given how subjective credences are defined in terms of fair betting odds, it is in fact plain *incoherent* to have credences significantly different from what one takes as the Humean objective chance. If all this is correct, it shows that Humean chances definitely are apt to play the PP role of governing credence—something no other account of chance has ever achieved. But our discussion has left open a number of questions; I will address three such questions briefly now.

4.1.4.1. From *n*-Case to Single-Case Reasonability?

The pragmatic argument looks at a situation in which we need to guess the outcomes of S setups over a decent-sized number of "trials." But sometimes, of course, we only need to make an educated guess as to the outcome of a single "trial"—e.g., if you and I decide to bet on the next roll of the dice, and then quit. If it works, the a priori argument steps in here and assures us that, even considering just one instance of a chance setup, it is incoherent not to have credences (nearly) identical to the objective chances, when the latter are known and the conditions of applicability of PP are otherwise met. But it is interesting to look at whether the pragmatic argument by itself is enough to justify application of PP in "single-case" circumstances. Assuming, as always, that we have no inadmissible information regarding this upcoming single case, the answer is "yes, it is." The argument shows that setting our level of credence in A to the objective chance is a "winning" strategy most of the time when guessing many outcomes is undertaken. But setting our credence for outcome A equal to a *constant*, x, over a series of n trials is *the same thing as* setting our credence equal to x for each individual trial in the collection. It cannot be the case that the former is reasonable and justified, but the latter—its identical sub-component—is unreasonable or unjustified. Just as two wrongs do not make a right, concatenating pragmatically unreasonable acts does not somehow make a composite act that *is* reasonable. So when we have to set our credence—make a bet, in other words—on a single case, the prescription of PP remains valid and justified by the pragmatic argument, even though that justification is, so to speak, inherited from the fact that the PP-recommended credences are guaranteed to be winners most of the time, over medium-run sets of "trials."

Suppose the contrary. That is, suppose that we think it is possible that over decent-sized numbers of trials using PP is justified, but nevertheless in certain *specific* single cases—say, *your very next* coin flip—it may in fact not be justified. What could make this be the case? If we accept TPI, there has to be some ground for this asymmetry. For example, you might think that some specific local factors exist that make a higher-than-0.5 credence in heads reasonable. Perhaps you have put the coin heads-up on your thumb, and as it happens, your coin flips tend to turn over an even number of times, so that flips starting heads-up (tails-up) land heads (tails) more often than 50% of the time. What this amounts to, of course, is postulating that *your* coin flips in fact comprise a different SNM than ordinary coin-flipping devices (including persons)—condition 5 in our earlier (chapter 3) example of the fair coin-flipping SNM is violated by these assumptions. Nevertheless, let's assume for the moment that the 50/50 chance of heads and tails on coin flippings is in fact a proper part of the Best System, and that your flips do fall under the 50/50 regularity (perhaps the Best System is content with a loose specification of the coin-flip setup). By our assumptions, your flips *also* satisfy the conditions for a less common SNM. Now there are two cases to consider: first, that we (who have to set our credences in *your next flip*) know about this further objective chance; second, that we don't.

If we do know about it, then the case is parallel to the breast cancer example from chapter 3. Knowing the higher-than-50% chance for your flips when they start heads-up, and seeing that indeed you are about to flip the coin starting from that position, we should set our credences to the higher level. The rules of admissibility tell us that the ordinary 50/50 chances are inapplicable here, but the chances arising from the more constrained SNM that your flips instantiate are OK. This does nothing, of course, to undercut the reasonableness of PP in *ordinary* applications to a single case, when such trumping information is not available. Chances are guides to life *when you can't find a better one*; sometimes the better guide is a different, more specific objective chance.

If we do not know about the higher-than-50% chance, then the question becomes: is setting our credence to 0.5 as PP recommends reasonable? It might seem that the answer is negative, because the "real" probability of heads on that next flip is in fact higher. But as Humeans about chance, we know that this way of thinking is a mistake. Both objective chances are "real"; the fact that sometimes one trumps the other does not make it more real.[14] An application of PP is not made unreasonable just because there is a *better* way of guiding credence out there, inaccessible to us; it only becomes unreasonable if that better way comes to the chance user's attention, and when that happens, the original chance can no longer be plugged into PP because of the violation of admissibility.

The case here is conceptually the same as that of the crystal ball that reliably shows the future. We already have rejected the idea that an event should be held to be "not a chance event" if it is possible that a reliable prediction can be made concerning its outcome, using tools ordinarily unavailable to us (like crystal balls, or Laplacean demons in a deterministic world). Finding out about the higher-than-50%-personal-chance of heads is just like finding a crystal ball showing the next flip's outcome, only less reliable. The mere *existence* of either one is not grounds for saying that the ordinary 50/50 chance does not cover your coin flips.

4.1.4.2. Mr. Unlucky

It might seem that both the pragmatic and a priori arguments must fall short of demonstrating the irrationality of not adhering to PP.

---

[14] And as we will see in chapter 5, it is possible for the more generic, macro-chance to trump the more specific, micro-based chance.

For it seems conceivable that someone could accept all the premises of the arguments, and yet stubbornly insist that *for him*, PP is neither a good strategy nor rationally required. Mr. Unlucky, for example, knows all about HOC, but thinks that the chances are of no use to *him*, because he is reliably unlucky. Those rare deviations of the frequencies from the chances? Mr. Unlucky is sure he spends most of his life in their vicinity. He does not know *how* events around him will prove to be anomalous and deviant from the perspective of the HOCs; he is just sure that they will be so. Does either of our arguments for PP suffice to convict Mr. Unlucky of irrationality?

I do not see that they do. Mr. U may or may not be correct about the anomalous character of events around him, but *given* his convictions, we can hardly convince him that betting according to PP's advice will be the best strategy he can follow. (It may not look any *worse* to him than any other strategy—he may believe that any strategy he adopts is sure to fail.) Nor can we convict him of violating the principle of non-contradiction or even TPI. Mr. U thinks he has a *reason* to discriminate between certain events in setup *S* and others: some of them occur around *him*, and therefore are prone to anomalous results in the medium-run.

But seeing why Mr. U's beliefs do not violate TPI or PP also points to the reason why Mr. U need not be seen as undermining our deduction of PP. For Mr. U, we now see, takes himself to have inadmissible information concerning the chancy events in his vicinity: namely, their proximity to himself, or perhaps his interaction with them. Mr. U is the kind of person who thinks that his betting on a sure-thing horse at the track is enough to dramatically lower the likelihood of its winning. But then knowing of (or positing) a bet by himself constitutes information relevant to whether the horse wins or not, and such information is not objective-chance information;[15] therefore it is inadmissible, and Mr. U is rationally free to ignore PP.

---

[15] At least, it is not *HOC*-chance information, which is what matters here. Mr. U may take his bet to increase a "propensity" of the horse to lose—exactly how he thinks of his bad luck as interacting (or not) with the world is irrelevant to the admissibility issue.

Suppose we rework our description of Mr. U so as to carefully strip away any hint that he takes himself to have inadmissible information. This requires deleting any opinion about events in his vicinity being anomalous, or he himself being unlucky, etc. A stripped-down, purified Mr. U will simply insist on having credences other than the chances for certain events or sets of events, *for no reason in particular.* We can surely imagine such an agent. But now the a priori argument comes back into play—such an agent clearly violates TPI. And equally importantly, I think we all share the intuition that such an agent is being irrational. So Mr. U does not give us a counterexample to our deduction of PP, after all.

4.1.4.3. Few- and No-Case Chances

The Best System aspect of Humean chance takes us away from a (sophisticated) actual frequentism in two ways: by "smoothing out and rounding off" the chances and chance laws, and by using higher-level and lower-level regularities to extend the domain of objective chances to cover setups that have few, or no, instances in actual history (like our 43-slot roulette wheel). Both aspects make objective chances easier to discover and to work with than pure finite frequencies, which is all to the good in a concept, such as the concept of objective chance, whose nature is bound up with its utility to finite rational agents. Since it never drags the objective chance far away from the frequencies when the numbers of actual instances of the setup in history are large, it is not problematic vis à vis the deduction of PP. But the latter aspect may well look problematic. For we know that if the number of actual instances of setup S in all history is relatively small, then the actual frequency may well be quite far from the objective chance dictated by the higher-level pattern. Perhaps 00 lands in $\frac{1}{25}$ of the times (say, only a few hundred) that our 43-slot roulette wheel is spun, rather than something close to $\frac{1}{43}$. In cases such as this, how can PP be justifiable?

Notice that in cases like this, where the numbers over all history are relatively low, we cannot mount an argument similar to our basic argument for PP in the preceding, but now in favor of a rule of setting credence equal to the actual frequency. Why not? Because there is no guarantee that there will be the sort of uniformity over subsets of n

trials that we knew we could appeal to when the numbers are large or infinite. Suppose our 43-slot roulette wheel was spun a total of 800 times, and 00 came up 32 times. Considering now guesses as to the number of 00 outcomes in "short run" sets of $n = 50$ consecutive spins, can we assert with confidence that in most of these runs, a guess of $\frac{2}{50}$ will be closer to the actual frequency more often than a guess of $\frac{1}{43}$? By no means. It may happen that the actual pattern of outcomes is very much as one would expect based on the objective chance of $\frac{1}{43}$, but in the last 90 spins 00 turns up a surprising 12 times. These things happen, and given the low overall numbers, cannot be considered as undermining outcomes for the objective chance of $\frac{1}{43}$.[16] But if this is how the 00 outcomes are distributed, then a person betting on a subset of 50 consecutive spins will do better to bet with the objective chance, *most of the time* (just not in the last 100 spins!). Neither a credence level of $\frac{1}{25}$ nor one of $\frac{1}{43}$ is guaranteed to be a winner in most of the short-run guessing games that might be extracted from the total 800 spins.

It is true, of course, that if we consider all the possible ways of selecting sets of 50 outcomes for our guessing games—not just consecutive spins, but even-numbered spins, "randomly" chosen sets, etc.—then the frequency of $\frac{1}{25}$ is going to do better, overall, in this wider set of games. (The reason is that in the vast range of games where the 50-trial subsets are chosen "randomly," the frequency in such subsets will most often be close to $\frac{1}{25}$ rather than $\frac{1}{43}$.) But it remains true that a consequentialist argument for setting credence equal to the actual frequency rather than the HOC is much weaker here than is the argument for using HOCs in the standard case of setups with very many outcomes. And we should remember that if the divergence is too serious, and the number of outcomes at issue large enough, the Best System will have to go with the frequency rather than the symmetry-derived chance: we have frequency-deviation tolerance, but only within limits.

The credence level of $\frac{1}{25}$ *is* guaranteed to beat that of $\frac{1}{43}$, of course, in the limit as the "short" run over which we are guessing approaches,

---

[16] Chapter 5 will contain a full explanation of and discussion of undermining.

and finally equals, the total set of 800 actual spins. So if one somehow knew the actual frequency of 00 over all history, *and* one somehow knew that the 800 spins on which one was betting constitute the entirety of that history, then one would be irrational to set one's credence equal to the HOC. The actual frequency would be a better credence level. But clearly, these two facts would jointly constitute inadmissible information, so PP would not be applicable in this scenario. Equally clearly, without a crystal ball we have no way of getting hold of this kind of information before the fact. So in this scenario, we are stuck with the following situation: An ordinary agent familiar with the nature of HOC will (correctly) suppose that the chance of 00 is $\frac{1}{43}$, set his credence accordingly, and if he continues betting long enough at the odds given by his credences, he is bound to lose (if betting *against* 00). That is unfortunate, but not something that undercuts the pragmatic argument for PP overall. In a chancy world, when one's information is all admissible, there can be no across-the-board guarantees of success.

4.1.5. Hall on the Humean's Chances of Deriving PP

Recall that Ned Hall (2004) argues that a Humean reductionist about chances faces insuperable obstacles to any attempt to demonstrate the rationality of PP. His main argument (sec. V) is based on a few-case scenario, similar in some ways to the $\frac{1}{43}$ roulette wheel, but even closer to the toy example discussed in chapter 5, section 5.1. The key to his objection is to posit a scenario in which one *knows* that only a few trials of a certain chance setup will ever be actual. In some such scenarios, intuition tells us that setting our credences using PP under HOC is unjustified. As we will see in chapter 5, the correct response to this sort of case is to agree that PP is not justified, either because one has inadmissible evidence or because it is an application of HOC that goes beyond the natural, intended limitations of HOC (or both). Which one is the right thing to say depends on the details of the case. Hall takes the existence of such cases as an insuperable obstacle to the reductionist's hopes of justifying PP, but in fact it just reminds us that care is needed in understanding both admissibility and the proper domain of application of HOC.

4.2. PP and the Epistemology of HOC

4.2.1. Strevens on PP and Induction

Let's return briefly to the impossibility arguments of Strevens (1999). He compares the impossibility of proving PP with the impossibility of solving Hume's riddle of induction, or at least one version of it: ". . . there is a resemblance between rules for probability coordination and the rule of enumerative induction and a corresponding resemblance in the difficulties encountered in pursuit of their justification" (pp. 248–249). And in certain ways, certain attempts to justify PP do run into the sorts of problems that block attempts to solve Hume's riddle of induction. But solving Hume's riddle is not part of my aims; rather, I help myself to our usual inductive practices and assumptions. Without induction, there would be no point in arguing for a Humean theory of chance: from the stability of patterns and frequencies we have observed to date, we could infer nothing about whether the future will be similar, and hence nothing about whether there exist objective chances for processes not yet complete.[17] With induction, on the other hand, we can presume to have a good idea of what the actual facts about patterns of events in our world look like—i.e., we more or less know the base facts that determine the Humean chances. So when it comes to justifying PP, far from being in a position similar to the philosopher attempting to solve Hume's riddle, instead our position is a very good one. Trying to prove the reasonableness of PP *given* HOC is like trying to prove

---

[17] The case is different if we presusppose a growing block view of the world and (hence) the Humean Mosaic; the chances supervene on just past history, and hence exist *now* whether or not the world in the future suddenly starts behaving quite differently. On the other hand, justifying the PP then becomes more tricky, as we will see in the next subsection.

the rationality of induction in science, *given that there are universal laws of nature*—that is, dead easy.

As we saw earlier, Strevens anticipates and discusses views that build in frequency guarantees by definition (though not anything quite like our HOC), and argues that nothing real is gained. His response is something along the lines of: "OK, you've just pushed the bulge under the carpet to a different place. Now your problem is explaining to me why I should ever believe that there exist such laws/chances, and how I can come to be confident I know them."

However correct a reply along these lines may be to the attempt to justify induction using laws of nature, it is toothless against HOC. I can rely on ordinary induction as the answer to the question of how we know that HOCs exist and how we come to know their values. Presupposing that Hume's riddle can be set aside, I have a way of getting my hands on the existence and values of HOCs: ordinary methods of observation and scientific inference. Chance setups can be identified by a variety of methods; stable outcome patterns can be discerned and verified by standard statistical practices; aptness for inclusion in the Best System can be checked by means both theoretical/conceptual and empirical; and most importantly, worries about whether chance setups behave radically differently in other regions of the universe can simply be set aside. If they do, then the Humean chances are different in those regions, full stop, and that fact does nothing to undermine the existence or utility of our local chances, here and (cosmically speaking) "now." The pragmatic, user-friendly HOC that emerges from chapter 3 makes clear that, while there may be controversial borderline cases and chances whose values are hard to determine empirically, by and large there is no great epistemic barrier to knowing the chances.

Notice that the very use of statistical inference from data, if understood from a Bayesian perspective, involves invocation of the PP to justify the crucial step from observed frequencies to confidence in the objective chances having certain values. But there is no problematic circularity. We use the PP (plus ordinary induction) to arrive at confidence about what objective chances there are, and what their values are. Is PP justified for HOC? Yes; and the arguments invoke only the general nature of Humean chance, not any specific objective chance rules inferred from experience. So the PP is not used to justify itself.

We do need to assume that ordinary induction, i.e., assuming that the future will be in most ways similar to the past, is reasonable, in order for the epistemology of HOC to be reasonable. But the same holds for many philosophical views and projects, and of course for our beliefs in both science and daily life. As we will now see, HOC might have to rely in an even stronger way on induction, if we adopt an A-series view of the ontology of time.

4.2.2. Justifying PP in a Growing Block World

Both justifications of the PP for HOC given in the preceding worked implicitly in a framework in which rational agents have to make bets concerning heretofore-unseen instances of the instantiation of a chance setup S. In the a priori argument, those instances are stipulated to be *future* instances. And in both arguments, appeal is made to what the HM must be like, over *all* history, for the objective chances to be what they are. So it is clear that facts about future chance events play an important role in the justifications of PP. With our standing assumption that the HM is the "block universe" of eternalism, this is no problem. But what can we say if there are no facts about future events, and HOC's chances must supervene on only the past history of our world? It seems to me that in this scenario the justifications of PP themselves need to build in an assumption of the reliability of induction, and the resulting arguments are importantly different.

The idealized scenario for invocation of PP may be thought of like this: Suppose God tells you that the objective chance (HOC) of A in S is x, and you have no inadmissible information relevant to whether or not A. Then your credence in A, if you have one, should be x. But in a growing block universe you know that the HOC rule for S supervenes only on world history up to now; *logically* speaking, what God tells you implies nothing about what the future will be like. Now suppose that you are nonetheless committed to induction, to assuming that (in the relevant ways we all implicitly understand) the future will be much like the past. One then trusts that, when the universe has grown and aged some more, the chance rule for S that God revealed will still hold. One trusts, therefore, that those who adjust their credences to the objective chances in the intervening time will have adopted an optimal strategy, in the sense sketched in the consequentialist deduction of PP. Thus, adjusting one's credences to the God-revealed *current* objective chances in S is a justifiable strategy when considering how to bet on the *next* or *next several* instances of S.

This shows that the consequentialist deduction of PP retains its basic force, weakened only insofar as one has doubts about the reasonability of applying induction to Humean chances. But what about the a priori deduction?

Here it seems to me that the argument is weakened in a more substantial way, though not completely undone. The strength of the a priori argument lay in the fact that it is literally, logically incoherent to believe that bets at odds other than those corresponding to the Humean chances are fair, *in an indefinitely long sequence of bets on future instantiations of S.* The incoherence is due to the fact that one knows that if there is such an indefinitely long future sequence, the frequencies found in that sequence must match the *current* Humean chances (i.e., the chances God has whispered in one's ear), given that the latter supervene on all history, including that indefinitely long future sequence. If, by contrast, the current Humean chances God reveals supervene only on events up to now, it is not directly incoherent to suppose that bets could be fair at different odds, in that indefinitely long future sequence.

It is, however, incoherent to have a policy of, in general, expecting the future to be like the past in all the important nomological and quasi-nomological respects to which we usually apply induction; to have no particular reason to think that the objective chance rule for S will change in the future; and to nonetheless expect the indefinitely long sequence to have frequencies that would entail such a change in the S chance rule. So it seems we can conclude that agents who are committed to ordinary induction when forming their expectations about the future would be incoherent if they violate PP.

Now let's look again at the non-circularity mentioned at the end of section 4.2.1. If we have to presuppose induction in order to justify

PP, it looks like we may have trouble, because PP itself is used in the standard inductive practices of (Bayesian) statistical inference. Do we have here a vicious circle at last?

I believe the answer is still "no." Induction is presupposed in justifying PP for HOC in a growing block universe; but as we saw, it is the generic "future-will-resemble-past-in-important-respects" form of induction that is presupposed. When PP is invoked in an inference from statistics to chances, while the result is "induction" in some sense (that is, ampliative inference), it is not at all the same sort of ampliative inference. In fact, it is an inference only to facts that supervene on *past, unobserved* events, not future ones! Since in a growing block the Humean chances supervene only on history-to-date, the standard statistical inference from data showing frequency x for A in S to the chance of A in S being, with high confidence, near x, is an inference taking us from existing (past/present) facts to other existing (past/present) facts. This inference does not rely on the generic, future-looking principle of induction that we had to invoke in order to make our justifications of PP go through. So there is, it seems, no worrisome circularity here either. That said, it is clear that things are simpler and easier for the Humean if she embraces an eternalist ontology.

4.3. Other Accounts and PP

Whatever the limitations of our deductions of PP may be, it should be apparent already that some of the standard competing accounts of objective chance cannot offer anything nearly as good. Still, it may be useful to examine how the competing accounts fare on justifying PP.

4.3.1. Hypothetical Frequentism

Before Howson and Urbach's argument, at least, the traditional hypothetical frequentist, like the single-case propensity theorist, had a position that utterly disconnected the objective chances from frequencies (and even from random-lookingness) in actual trials, so she could not mount an argument for the rationality of PP based on its guaranteeing a winning strategy most of the time (for us, in finite human history). Instead, she was forced to go second-order and say that adopting PP will yield winning strategies *with high probability*, this probability being an objective one derived from the first-order objective chances themselves. But even granting this second-order objective probability, the question remains: why should I *believe* that adapting my credences to the objective chances is a likely-winning strategy, just because this proposition has a high objective chance in the hypothetical frequentist's sense? Evidently, I am being asked to apply PP to her second-order chances, in order to establish that PP is justified for her first-order chances. If I reflect on the literal meaning of these second-order chances, they direct me to contemplate the limiting frequency of cases (worlds?) in which applying PP to the first-order chances is a winning strategy, in a hypothetical infinite sequence of "trial-worlds." The metaphysics begins to look excessive, and in any case we immediately see that the problem reappears at the second level, so that we need a third-order argument to justify PP for the second-order chances—and so on. Infinity turns out to be an unhappy place to mount a consequentialist argument for PP.[18]

With the Howson and Urbach argument, things looked a bit better for the hypothetical frequentist, but only if she could somehow ignore basic physical facts, such as the fact that coins wear out when flipped indefinitely, and that the universe itself will either collapse in a Big Crunch, or die an ignoble "heat death," both of which make literally infinite sequences of flips physically impossible. The Howson and Urbach argument relies on beliefs about literally infinite (though counterfactual) sequences of chance outcomes, since only in the infinite limit can a sure loss be derived for betting odds (hence credences) different from the objective chance. Now the usual response to a point like this is to brush it aside, noting that the infinity of a von Mises collective has always been put forth as an idealization, not something to be taken literally. But in this context, the infinity can't simply be

---

[18] The Howson and Urbach justification of PP examined in section 1.3 of chapter 1 was not, of course, consequentialist.

brushed aside: in long-but-finite sequences, according to the hypothetical frequentist's doctrines, the frequencies may be arbitrarily different from the true chances; hence there is no contradiction *at all* in violating PP, for von Mises chances-as-idealizations.[19]

The problem is fixable: one simply has to jettison the unlimited frequency tolerance for finite (but very long) initial sequences of chance events. Then one can prove an incoherence in having the belief that the objective chance of A in S is x, and having credence in outcome A x', where x' is non-trivially different from x. But to jettison the frequency tolerance, while still retaining a requirement of something like von Mises randomness (but adapted to finite segments of collectives), is to go two big steps in the direction of Humean objective chance.

4.3.2. Actual Frequentism

Both the consequentialist and the a priori arguments for PP under HOC relied on the fact that Humean chances are, in a sense, just modified actual frequencies. So it may seem plausible that an unretrenched actual frequentist could avail himself of these arguments too. To some extent this is correct, but in several ways the sophistications of HOC make a big difference, in favor of HOC.

One sophistication was the imposition of a requirement that the distribution of outcomes be nicely random-looking, over at least most of the overall pattern. Now in fact nobody who advocates an actual frequency account of probability would fail to impose such a

---

[19] There is a sense in which this may be unfair to von Mises. The hypothetical infinite collective's frequencies, he might reply, can be assumed to be the same as what we have seen in our finite samples so far. The collective is *our* creation, our idealization, after all, so we decide what its properties are!

This response works for a kind of hypothetical frequentism that explicitly uses the collective as a mere idealized model of the *actual* frequencies, and makes no claim that there is any fact of the matter about what *would* result if we *could*, counterfactually, extend the sequence of trials indefinitely. Such a view is compatible with Humeanism about chance, and might (if more fully articulated) turn out to be similar to HOC. By contrast, a hypothetical frequentism that claims that the objective probability just *is* the limiting relative frequency in a counterfactual infinite sequence of trials is vulnerable to this objection that a finite initial segment may have any frequencies you like.

requirement, in practice if not in theory. It's easy to see why. Suppose that the actual pattern of coin flips in the world consisted of 1,000,000 flips, the first 500,000 landing heads and the last 500,000 landing tails. The actual-frequency chance is then 0.5 for both heads and tails. But anyone living in such a world will do rather badly in her bets if she applies PP faithfully, compared to someone with a bit of common sense. She will obstinately accept 1:1 bets on tails all her life, and never once see a tails outcome.[20] By contrast, a normal person—even if told that the overall frequency of heads and tails is 50% over all history—will make a few bets at 1:1, start to notice that she is losing every time she bets on tails, and soon start to accept only much better odds if asked to bet on tails. After a few hundred, she will cease to take any bets at all on tails, and be very willing to bet on heads, at almost any odds. And in practical terms, she will be much better off than the obstinate finite frequentist who stays faithful to PP. So will the believer in HOC, since her system will have a "chance law" assigning $Pr(H|Flip) = 1$ for part of history, and $Pr(T|Flip) = 1$ for the rest. So, if somehow told the objective chance rule, she will not even go through the short but painful learning-curve experience of the normal person just described.

It may seem that the bad news for actual frequentism ends there, however. For when it comes to cases like the $\frac{1}{43}$ roulette wheel that in fact lands 00 $\frac{1}{25}$ of the time, it seems clear that the advantage on the pragmatic side goes to the frequentist. If the house offers close to 1:42 odds, and she bets on 00 long enough, she is bound to come out a winner. The Humean will see no reason to settle on 00, however, and may or may not do well, depending on where she places her bets. In the limit of single-instance setups, someone apprised of the frequency-facts can't go wrong, while the person who knows the HOCs can and often will.

All of this is, however, mostly beside the point, since these are cases where, absent divine intervention, no one can come to know the actual frequencies with any certainty. (In the all-heads-then-all-tails

---

[20] I'm assuming all her life occurs in the portion of history spanned by the first 500,000 flips.

world, one could not arrive at the HOC's either, until some time after the turnaround point.) Moreover, the pragmatic advantage that would accrue to knowing the actual frequencies, in certain of these cases, does not necessarily undercut the claim of HOC to be able to rationalize PP. For in such cases, knowing the actual frequency is the same as having inadmissible information, vis à vis application of PP to the Humean chances.

This can be illustrated nicely with one of the test cases first used by Lewis in his 1980 *Subjectivist's Guide*. One believes that the objective chance of a certain coin landing heads, when flipped, is 0.5. Most of Lewis' test questions ask: What should one's credence $Cr(heads|E)$ be, for various kinds of E? One case in particular is this: you know that the coin toss you are being asked about is one of 10 that occurred earlier in the day, 9 of which laned heads. Yet you remain convinced that the *chance* of it landing heads, on each flip of those 10, was 0.5. What should your credence be, for the particular toss just before noon (one of that group)? Not 0.5, Lewis replies, but rather something much closer to 1.0 than to 0.5. Having this frequency information is, Lewis correctly says, to possess *inadmissible* information vis à vis application of PP. Knowing actual frequencies, when possible, is useful; but it is not the same as knowing the objective chance, nor does its utility somehow undercut the claim of HOC to be our "guide to life" when better information is lacking.

What about actual frequentism and the a priori argument for PP? It may seem as though the argument goes through perfectly well for actual frequencies, but in fact we need to distinguish two distinct scenarios. First, suppose that the number of instantiations of the chance setup occurring *outside* the hypothesized indefinitely long sequence of trials is finite and, at some point during that sequence, becomes *much smaller* than the number of trials in the sequence. In this case, the argument works about as well as it does for HOC: the actual frequency in the long sequence *might* differ *slightly* from the objective chance (= overall actual frequency), but not greatly.[21] Second,

---

[21] Here we should recall that actual frequency chances are only well-defined if the total number of instantiations of the setup or reference class is finite. However many past instantiations there have been, the *indefinitely* many (but still always finite!) future instantiations will eventually wash out their influence on the final actual frequencies.

suppose that although over time the indefinitely long sequence keeps growing (never becoming infinite inside the world's history), so does the number of instantiations of the chance setup that occur outside of the sequence, so that the frequency inside the sequence is by no means able to dominate and "wash out" whatever the frequency outside the sequence may be. In this case, the argument fails to go through! There is no contradiction in supposing that the frequency in the sequence may differ from the overall actual frequency, even if the sequence is as long as you like (but still finite). The reason for this is one of the Achilles' heels of actual frequentism when compared to HOC: officially, it imposes no constraint on the distribution, in space and time, of the outcomes.

There is a way for actual frequentism to avoid this problem: restrict the domain of the setup whose objective chance is in question to the very sequence posited in the *a priori* argument. For example, make the coin-flip chance at issue be *the chance of Heads using this very coin*, the coin that will only ever be flipped in the indefinitely long sequence and then destroyed. If we make this restriction, then the argument goes through, and does so without any slack at all. But it is a somewhat pyhrric victory, since it does not extend to chances for type of events that we must view as occurring out in the wild, and not just inside a hypothetical long-run experiment; which is to say, almost all chances that we might be interested in.

Seeing how the *a priori* justification of PP fails for actual frequentism illustrates once again how core features of HOC are crucial for capturing our concept of objective chance. The ideas of random-lookingness and unpredictability are quite central parts of that concept, and these are not found in the notion of an actual frequency. Our intuition that AFs can be quite far from the "true" objective chances if the total number of instances is low is also important, and is captured by HOC in an elegant way. And finally, of course, it also matters that actual frequencies are less epistemically accessible than HOCs. Where we have reasons to believe that a setup is accurately thought of as an SNM, or is dictated directly by a physical theory (like QM), we have resources for inferring or calculating the objective chances that go well beyond mere frequency data. Not all objective chances are like this, but many of the most central cases, from card games to physics experiments, are. And for these cases,

HOC makes our actual epistemic practices much more intelligible than simple actual frequentism can.

4.3.3. Propensities

The metaphysical propensity theorist, by not offering any substantive or reductive definition of objective chances, would appear to have little or nothing to say about why PP is rational to apply to our beliefs about propensity-chances. This does not necessarily stop them from claiming the high ground. The boldest version of this position in recent times appears in Hall (2004).

Let's recall the full force of Lewis' challenge to the advocate of metaphysical propensities:

> Be my guest—posit all the primitive unHumean whatnots you like. (I only ask that your alleged truths should supervene on being). But play fair in naming your whatnots. Don't call any alleged feature of reality "chance" unless you've already shown that you have something, knowledge of which could constrain rational credence. (1994, p. 484)

As we saw in chapter 1, Hall answers Lewis' challenge thus:

> What I "have" are *objective chances*, which, I assert, are simply an extra ingredient of metaphysical reality, not to be "reduced" to anything else. What's more, it is just a basic conceptual truth about chance that it is the sort of thing, knowledge of which constrains rational credence. Of course, it is a further question—one that certainly cannot be answered definitively from the armchair—whether nature provides anything that answers to our concept of chance. But we have long since learned to live with such modest sceptical worries. And at any rate, they are irrelevant to the question at issue, which is whether my primitivist position can "rationalize" the Principal Principle. The answer, manifestly, is that it can: for, again, it is part of my position that this principle states an analytic truth about chance and rational credence. (2004, p. 106)

In one sense, Hall seems clearly to decline to "play fair" and *show* why PP holds for his primitive chances. But in another, strictly technical, sense, Hall does abide by Lewis' injunction. He makes a terminological stipulation that "chance" is to be by definition something about which PP is analytically true, so it follows trivially that knowledge of these primitive chances constrains rational credence. But as we observed in chapter 1, with the same form of argument one can prove that the laws of nature can be known a priori to hold in the future. Or, again with a similar argument, we can prove the immortality of the *soul*, if we wish.

But of course, knowing that souls are immortal is unlikely to comfort us near life's end, until we are sure that such things exist—the skeptical worry is hardly "modest"! The same is true for propensity chances as stipulatively defined by Hall. So until the chance-primitivist can overcome the modest skeptical worry that, perhaps, there are no such primitive chances in our world after all—and for Hall, as for Mellor (1995), Lewis (1994), and others, this means *proving determinism false*—we may as well set aside propensities and stick to Humean objective chances, whose existence is guaranteed and which can be learned by ordinary scientific practices.[22]

4.3.4. No-Theory Theory

Finally, let's consider Sober's (2005) no-theory theory of chance, briefly seen in Box 1.1 in chapter 1. It has the great virtue, compared to metaphysical propensities and hypothetical frequentism, of letting us be

---

[22] Hall goes on to suggest that perhaps the primitivist will be able to argue that the reasonableness of PP follows from constraints on reasonable initial credence functions that are connected to "categorical" features of the world. But although Hall speaks of these constraints as if they were "imposed by" categorical facts (2004, p. 107), in fact the constraints he has in mind are just rules about how to distribute credence in light of the presence or absence of certain categorical features of things (e.g., Hall suggests the constraints might include ". . . various indifference principles, carefully qualified so as to avoid inconsistency" (p. 107)). He goes on to postulate a situation in which exchangeability works as a "categorical constraint," allowing a hypothetical frequentist to derive the appropriate local application of PP. The problem of course is that exchangeability is by no means imposed by categorical facts in the world, and Hall's application of it here is in reality just a disguised application of PP itself. Why should my credences satisfy exchangeability? Because I believe there is a unique, stable objective chance for each trial, and I want my credences to match it.

sure that the chances exist and are non-trivial, because the chances just *are* whatever our accepted scientific theories say they are. It also beats those two competitors when it comes to justifying PP. Whereas they can give, in the end, no justification of the reasonableness of PP at all, it seems to me that Sober can appeal to our successes in using objective probabilities as *inductive evidence* that applying PP is a good strategy.

Or rather, Sober can help himself to such inductive support in sciences where the chances are not inferred from statistics, but rather grounded *a prioristically* in theory (i.e., in QM and statistical mechanics).[23] As we saw in sections 1.3 (chapter 1) and 4.2.1, the standard procedures of inferring objective probabilities from statistical evidence *use* PP—that is, their claim to methodological soundness rests on assuming that the objective chances being guessed at deserve to govern credence. Where PP needs to be assumed in arriving at the (estimated) chances, it would seem to be circular to claim that the success of the sciences using said chances argues for the validity of PP as applied to chances in that science. This caveat substantially reduces the range of objective probabilities for which the no-theory theory can claim an inductive grounding for PP.

So the no-theory theory may do better than primitive propensity views, but it only enjoys this advantage in sciences where chances are derived from pure theory.

4.4. Summing Up

There are two ways to demonstrate that if objective chance is as the theory of HOC claims it is, then rationality requires agents to respect the PP, or at least to come extremely close to respecting it. The wiggle room left open for rational agents is in any practical setting insignificant, so we can justifiably regard the PP as being demonstrated for HOC. This is a key advantage of HOC in comparison to other accounts of objective chance, all of which face more significant challenges in laying claim to the rational requiredness of the PP.

---

[23] These two cases are different from one another, and may require quite different treatments; see chapter 6. But in neither are the probabilities arrived at purely by statistical inference based on data.

5

Undermining

A large proportion of the discussion of Lewis' approach to objective chance has centered around the problem of *undermining* and the apparent contradictions that it may give rise to, when Humean chances and the Principal Principle (PP) come together. The issues arising in connection with undermining and PP are complex and thorny. But getting a good grasp on them is essential in two respects. First, showing that undermining is a soluble problem is crucial to any Humean approach to objective probability, for reasons that will shortly be clear. Second, more positively, treating the undermining problem will help us to clarify both the nature of PP, and the nature of HOC. This chapter will be devoted to explaining and resolving the undermining problem. Readers who already believe that the undermining problem is no problem for Humean accounts of chance should feel free to move on without wading through these waters, but for such readers I will mention that the upshot of everything is that PP requires a slight modification, one that in ordinary applications amounts to no change whatsoever. In Box 5.1 at the end of this chapter I'll consider whether the deductions of PP in chapter 4 serve to justify my replacement-principle NP*.

Shortly after first encountering Lewis' Humean account of chance, I wrote an article on the undermining problem (1997), and like many others at the time I argued against Lewis' solution. In particular, I argued that Lewis—and basically everyone else writing about undermining and PP at the time—was deeply mistaken about the scope of applicability of *Humean* objective chance. (Coming to this viewpoint was the first step on the path that led me to the present account of HOC.) In the next section I will sketch the undermining problem and recap my early conclusions about the scope of HOC. But we will see that, in light of the full account of HOC from chapters 3

and 4, the solution to undermining I advocated in 1997 is not acceptable. In section 5.5 I'll offer a better solution.

5.1. Undermining—What Is It?

Undermining is an inevitable feature of Humean theories of chance, if
the theory gives enough weight to actual frequency facts. When that
is the case, the objective probabilities supervening on the Humean
mosaic will entail that certain sorts of "unlikely" large-scale events
have a non-zero objective chance, even though, counterfactually, *were
said events to have occurred, then the Humean mosaic would have been
so different that the objective chances themselves would have been different*. For example, imagine that starting now, and for the rest of time,
radium atoms were to decay on average four times more rapidly than
they have up to date. That is, they still decay with statistics that fit
an exponential decay rule, but the half-life that one would calculate
based solely on future observations would be one-quarter of the half-
life observed in history up to now. Let F stand for the proposition
that future radium decays display such a pattern. If the total number
of radium atoms that exist in our world over all time is finite, then
this future not only is "physically possible" according to the accepted
(and let's assume, actually true) decay law, it even has a finite, non-
zero probability of occurring. That probability is *very nearly zero*, of
course, but it is not strictly zero. But plausibly, were such a future in
fact real, then the Humean Best System of chances would not include
our actual decay rule for radium atoms. Instead, there would be a new
law specifying the actual decay rule for the first 13 billion years or so
of time, up to now, and a different rule with a half-life one-quarter
as long for the rest of time.[1] F, in other words, would *undermine* the

---

[1] You might worry that the loss in simplicity engendered by such a modification of
quantum chances would be so bad as to outweigh the improvement in fit. I think this
is implausible. The trade-off is adding exactly one further chance-rule to the system,
in exchange for a vastly better fit with events that occur all over the place, in the later
epochs of the universe. But never mind: if this example does not convince, it is easy to
construct others by piling on further improbabilities (e.g., extending the half-life shift
to every radioactive substance), or simplifying the example by discussing a toy example

actual chances; if we assume that the chance rule for radium decay in our world *is* in fact as we think it is, then F cannot be true in our world. Yet F has a non-zero objective chance. Puzzling, to say the least.

The puzzle gets elevated to the status of a contradiction as follows: If F is our undermining future, then recall that PP2 says

$$C(F|H_{tw}T_w) = x = \Pr(F), \quad x > 0$$

where Pr is the objective chance function given by T_w, the Best System of chance for our world, and H_{tw} is the history of world w up to time t (the *whole* Humean mosaic prior to t). But as we have just seen, F plus H_{tw} jointly entail, under the Humean analysis of chance,[2] $\neg T_w$. So the standard axioms of probability tell you that $C(F|H_{tw}T_w) = 0$. Never mind that the difference between x and zero will be infinitesimally small, in any realistic case; never mind that neither H_{tw} nor T_w can ever really be known by us, and so forth: a contradiction is a contradiction, and is bad news.

5.2. Is the Contradiction Real?

A large number of authors have either questioned the legitimacy of the derivation of the contradiction, or proposed solutions to the problem different from Lewis'; to review them all would make this chapter unwieldy without shedding much light on the position I now advocate.[3] Here I will briefly discuss a refutation offered by Barry Loewer, and

---

(as, for example, when we discuss worlds that consist of nothing but a few particles, or balls that never interact but change colors randomly according to a chancy rule, and so forth).

[2] Here I take the Humean analysis of chance as an *a priori* truth, so that it automatically gets credence = 1 and can be added to the right hand side of the conditionalization without modifying anything. Strictly what a rational agent finds contradictory is not the set $\{F, H_{tw}, T_w\}$, but rather $\{F, H_{tw}, T_w, HOC\}$. If we suppose that HOC is not an *a priori* truth, the undermining contradiction still arises as long as the epistemic agent assigns some non-zero credence to the proposition *HOC*.

[3] The most recent contribution that I know of is (Belot, 2016), which argues that undermining is even worse for Humeans than they have realized; see (Hoefer 2018) for a response to Belot.

then a kind of response to the problem that seeks to make it go away by, in effect, replacing PP with something more general. This latter move is advocated, in distinct but related ways, by John Roberts (2001) and Jenann Ismael (2008). Their arguments will make clear what a recondite and non-central problem undermining is, though I don't think either fully succeeds in arguing the problem away. In the next two sections I'll develop my own way of eliminating the contradiction.

Loewer's (2004) response begins by noting that undermining propositions about future chance events, like our F, do not come into contradiction with the actual chances just by specifying a long series of outcomes. (Let's suppose that F simply gives a statistical description, at an appropriate coarse-grained level, of decay frequencies in the set of actual radium atoms existing at t, up to the much-later time t^*, at which point most radium atoms have decayed.) Rather, Loewer says, in order to make the contradiction arise we must further specify that *the posited series of outcomes constitutes the whole future* (for events of that setup type) in our universe. If that were not included in F, then F would not rule out an indefinitely long further-future after t^*, in which huge further numbers of radium atoms exist and decay in a way that fits the postulated original decay chance rule. If this post-t^* future is long enough, Loewer thinks, it renders the events in F a mere temporary statistical fluctuation, which is then *not* an underminer after all.

Call this further fact that Loewer says we need, specifying that F gives the more or less complete future in our world as far as radium atoms go, G. There *is* a contradiction between F, H_{tw}, T_w, and G. But G cannot be part of our background evidence about the world; and in any case, it is inadmissible with respect to F. So, Loewer claims, the contradiction never existed in the first place.

Loewer is right to point out that in general, our background evidence can never include certainty about how many total instantiations of a chance setup S there will be, in the future of our world. But I think he is wrong to view this as a complete resolution of the undermining problem. In certain special cases the contradiction can arise for a

Lewisian system without adding an extra, inadmissible premise. And it is even clearer that in a Humean system of chances such as I advocate, undermining does not require positing anything about the entire future history of the universe. Let's take these points in turn.

Loewer's own toy example for the undermining contradiction used a bare-bones world consisting entirely of coin flips. My preceding example used radium decays in our world. Loewer's F does not carry, on its own, the fact that the specified sequence is the entire future history of coin flips, whereas I first presented my F as implicitly containing the fact that the F-specified sequence of decays contains all the decays there are in the future history of our world. Then, in light of Loewer's objection, I broke it down into an unobjectionable F and the inadmissible G. Loewer is right to object that G is not something we can help ourselves to, being inadmissible (as well as the sort of thing we can never know). The trouble is that, while *this* undermining example may not work without a helping G, Lewis-style Best Systems can easily be imagined that do the trick without cheating. For example, suppose that a chancy physics specifies the post-collision trajectories of particles according to a chance law, with Newtonian motion in between collisions. If a certain set of outcomes of chancy collision-motions takes place, all particles will congregate into a small enough volume to undergo gravitational collapse, becoming a Newtonian black hole, in which all motion (hence collisions) dies out after a finite span of time. In effect, the world comes to an end. The Lewisian Best System for such a world could allow calculation of the probability of just such a set of collisions, and if the set's members are "improbable" enough according to the Best System, they could constitute an undermining future for the Best System itself. But in this scenario, F plus H_{tw} plus the laws T_w do entail the entire future of the universe. The contradiction is regained, as a conceptual possibility at least. In order to reject the derivation, Loewer would have to argue that in such a case T_w is inadmissible; but this is a result usually assumed to be disastrous for the applicability of PP to Humean chances.[4]

---

[4] Here I am ruling out "Space Invaders" (see Earman 1986), particles that suddenly appear in a Newtonian world, zooming in from infinity after time t, at exponentially decreasing velocities, to wind up leading a normal life—an unpredictable occurrence

Notice that Loewer's proposed resolution of the undermining contradiction problem turns on supposing that, if we *don't* specify that a certain undermining sequence F is the whole future history, then we cannot prove it to be in contradiction with history plus the Humean chances. If one leaves open the farther future, there may be enough future decays that *do* fit the chance law well enough to erase the undermining status of the sequence F.

It's not at all clear to me that this supposition is correct, even for the Lewisian approach to chance. If F covers all radium decays in a significant chunk of space and time, we should rather expect the Best System to trade simplicity for better fit, stipulating a special clause of the radium decay law that applies only to that region of spacetime. One does not have to know all of future history, to know that a certain widespread, enduring statistical fluke is going to be enough to undermine the posited chances. If we did not see that clearly enough in chapter 3, the deductions of PP in chapter 4 ought to have made it clear. Neither the pragmatic nor the *a priori* deduction will work, if statistical freaks of unlimited size are allowed (even though they are compensated by an even longer non-freaky future). In the case of my HOC, with its pragmatic and user-friendly nature, the point is still more obvious. If radium atoms now in existence decay on average four times faster than they should, for the next millennium on Earth, then the HOC Best System would assign a special decay rule to that millennium, in line with the specified decay pattern. No inadmissible knowledge of the future is required to reach this conclusion. So if the true radium decay law is in fact what we think it is, then this sort of F is, all by itself, an underminer—no extra G needed. It seems to me that Loewer's point shows us that the contradiction is harder set up (hence, arises less frequently in a sense) than one might have thought, but it fails to resolve the problem entirely.

Roberts (2001) and Ismael (2008) attack the undermining problem by questioning the fundamental status of PP. Recall that in one canonical form, PP is written: $Cr(F|XE) = x$, where F is an outcome-specifying proposition, E is the rational agent's (admissible) background information, and X says that the objective chance of F in the relevant chance setup is x. Roberts and Ismael point out that in real life we can never actually have evidence that provides *certainty* concerning the objective chances. $Cr(F|XE)$ is therefore an idealization, not something that corresponds to the epistemic situation of a real agent. Instead, what we actually have are opinions concerning what the objective chances might be, these opinions having associated subjective weights. What we really can apply is a more General Principal Principle (GPP):

$$\text{(GDP)} \quad Cr(A|E) = \sum_i Cr(X_i|E)x_i$$

(This is Roberts' formula.) GPP says that our subjective credences should equal the weighted average of the probabilities (x_i's) for A's occurrence, where the weights are given by our subjective credences in the various possible objective chance rules ($Cr(X_i|E)$'s). It therefore looks like a much more realistic representation of our typical epistemic situation regarding objective chances: not knowing what the true objective chances are, but having, based on experience in general, views about what chance rules are more likely and what rules are less likely to be true.

Ismael (2008) arrives at a nearly identical formula, which she called the General Recipe:

$$\text{(General Recipe)} \quad Cr(A) = \sum_i Cr(X_i)x_i$$

This differs from Roberts' GPP only in lacking the conditionalization on E, the agent's evidence. In her (2015) Ismael seems to indicate that it is all right to conditionalize on E, so long as E contains no information from the future, which makes General Recipe essentially equivalent to GPP. So let's continue by looking at how Roberts proposes to avoid the undermining problem.

The derivation of a contradiction via undermining depended on seeing an outright contradiction between an agent's "evidence" XE (which plays the same role as $H_{tw}T_w$ did in the preceding) and the undermining F. But Roberts points out that real agents can *never* have evidence that entails, with certainty, propositions like X or T_w. So for real agents the contradiction cannot arise. Real agents are instead in the situation of applying GPP, and the left-hand side of GPP need never be zero for undermining reasons.

GPP does not make undermining per se disappear. Among all the possible chance rules that we assign some credence, certain Fs will be underminers of those rules. So we may be inclined to think that the contradiction problem is not going to have gone away just by moving to GPP, but rather simply be hidden inside some of the terms in the sums GPP prescribes. Roberts goes on to consider this possibility. Rewriting the left-hand side of GPP using the total probability rule,

$$C(A|E) = \sum_i C(X_i|E) \cdot C(A|X_iE) \qquad (1)$$

yields

$$\sum_i C(X_i|E) \cdot C(A|X_iE) = \sum_i C(X_i|E) \cdot x_i \qquad (2)$$

The first part of the summations are identical on both sides. Moreover, due to undermining, we can (*pace* Loewer) set certain terms on the left side equal to zero. These will be the terms i for which the outcome A is an underminer of the chance specified by X_i. But the corresponding terms on the right side will be non-zero. So now suppose we could apply PP to each remaining non-zero term in the sum on the left side of (2), replacing $C(A|X_iE)$ with x_i. Then the sums on the left and right sides would be identical, except certain terms on the left get zeroed out—therefore, the equality fails. Contradiction is restored.[5]

But Roberts denies that applying PP to the non-zeroed terms on the left side of (2) is legitimate. His view is that PP, or rather GPP, can only be applied to evidence that real agents can have—namely, evidence that does not single out a single chance rule as correct.

---

based on the state of the world before t. If we allow Space Invaders, then Loewer could reply that even after black hole collapse of the pre-existing particles, a rich later history based on the careers of new Space Invader particles could swamp and render irrelevant the putatively undermining F. But the lack of Space Invaders might, plausibly, be an axiom of the laws of a Lewisian Best System that has as a part axioms corresponding to Newtonian particle mechanics, since without such an axiom the system is indeterministic and hence lacks predictive power.

[5] The equations here differ from those published in Roberts (2001) by including a simple correction, which he pointed out to me in correspondence.

In effect, Roberts claims that all instances of the basic PP are invalid, because they correspond to an impossible epistemic situation (knowing *XE*). So he argues that the substitution in the last step of the argument is not acceptable. On the other hand, Roberts accepts the zeroing out of terms that correspond to undermining. So how can the equality hold? Evidently, since certain terms get zeroed out, the others must get boosted a bit to compensate, so that the overall equality remains true. Roberts gives a nice plausibility argument for this boost of the non-undermining terms in the sum. But he advocates no particular formula for calculating the boost for the non-zero terms.

While I accept the Roberts/Ismael point that real agents' situations always correspond to GPP rather than PP, I reject Roberts' claim that GPP is the more fundamental principle. In fact, it seems to me that the plausibility of GPP is parasitic on the prior plausibility of PP itself. PP can be made intuitive by thinking of it as corresponding to a scenario in which God, or some near-omniscient time traveler from the far future, whispers in your ear what the objective chances for setup *S* are. You trust this entity completely. If you have no better information concerning whether outcome *A* in *S* will result, then what should your credence in *A* be? *x*—obviously. GPP makes sense because we don't have God or time travelers around to tell us the objective chances in our world; we have to make educated guesses based on our own evidence. But each term in the GPP sum gets its justification, so to speak, from the prior acceptance of PP. PP is the "ur-notion," GPP is its generalization to our imperfect epistemic situation.

So I don't want to follow Roberts in rejecting PP itself (or its application instances in the revised contradiction argument); as a conceptual matter, PP is not valid only when one's evidence is the kind of evidence the real world allows us.

Ismael might reject the substitution step in the preceding reasoning as well, because she too rejects the original PP, which in (2008) she expressed in PP2 form:

$$\text{PP}_{\text{orig}}: \qquad Cr(A|HT_w) = Ch_{Tw}(A)$$

Ismael finds a flaw in the conditionalization on T_w, which represents a certainty about the correct theory of chance for a world that real agents can never have. The General Recipe instead represents our epistemic situation correctly, and captures what was intuitive about PP all along. Unlike Roberts, however, Ismael does not concede that the undermining problem can be recovered in the context of GPP or General Recipe. Like Loewer, Ismael argues that there is no need to zero out any $Cr(A|X_iE)$, because as long as A merely states the occurrence of some future events, *but not that they constitute the* complete *future history*, then no contradiction is present. And if A not only specifies an intuitively undermining event but also asserts that it is all the rest of history, then A is the kind of proposition that chance rules do not cover in any case. Like me, Ismael rejects the idea that a Best System of chances should be able or required to assign probabilities to total histories of the world (*qua* totalities, described as such). Unlike me, however, Ismael has in mind the type of Humean approach to chance that makes it impossible (or at least implausible) that any finite event F can by itself trigger the undermining contradiction. Earlier I argued that for my pragmatic version of Humeanism, HOC, this way out is not available.[6]

For the moment, then, I will proceed on the assumption that the undermining contradiction problem is still in need of a solution. Loewer showed that it may be even harder than one first thought to find a chance outcome F that truly undermines the chance X without the help of inadmissible premises; but it seems clear that such outcomes *can* in fact be concocted, for plausible chance rules in worlds like ours. Roberts and Ismael correctly pointed out that the basic PP is not something we can ever be, in fact, in a position to apply. But for HOC, the undermining contradiction problem reappears in the context of the GPP, or in Ismael's General Recipe once it is written with conditionalization on the agent's evidence. The lesson of our calculations with GPP is this: where undermining outcomes like F are concerned, the Humean about chance has to set credence in F

[6] I thank Ismael for engaging in a lengthy and very useful correspondence about these issues during the final stages of writing of this chapter.

equal to zero conditional on certain chance rules X_i being correct. Therefore, in order for her credences to conform to GPP (as rewritten in (2)), certain terms in the sum must have a value that is "boosted" compared to what the basic PP dictates. How should this be done? What is the *right* way to resolve the problem with our credences posed by undermining?

5.3. The Lewis-Hall Solution

Lewis (1994), Hall (1994), and others have argued that the correct way to resolve the contradiction is by abandoning PP in favor of a corrected, New Principle,

$$NP: \quad C(A|H_{tw}T_w) = x = Pr(A|T_w) \,^7$$

NP differs from PP by equating credence in A to the probability of A *conditional on T_w,* i.e., *given that the Humean chances are those captured in T_w.* In ordinary non-undermining cases, the quantity $Pr(A|T_w)$ will, they maintain, be essentially identical to $Pr(A)$, thus explaining why PP seems such a good rule in ordinary circumstances. But in cases where undermining threatens, *NP* can get us out of the contradiction. Since (Lewis and Hall maintain) the past is both admissible and no longer chancy, at or later than t we have $Pr(H_{tw}|T_w) = 1$, and so $Pr(A|T_w) = Pr(A|H_{tw}T_w)$. But $Pr(A|H_{tw}T_w) = 0$, by the axioms of probability (as we saw on the credence side, earlier—here we are assuming A is an underminer, like F). So chance and credence have been restored to compatibility.

But at what a cost! In chapter 2, I noted that, contrary to what Lewis and others appear to take for granted when discussing Humean chance, we should not presume that the Best System will assign objective chances to all propositions, nor even to all propositions specifying a well-defined physical state of affairs. The Best System might only

---

[7] Hall sees this change not as a correction, but rather as the way we should have been thinking of PP all along. Reasons for disputing this notion will be seen shortly.

cover actual events in the HM in a patchy and piecemeal way. But for the Lewis-Hall NP to be well-defined, objective chance has to swallow a huge new domain: namely, itself.

For Lewis, recall, objective chances are time-indexed *unconditional* probabilities. So the conditional probability $P_{tw}(A|T_w)$ is calculated using the "definition" of conditional probability, and thus is equal to $P_{tw}(AT_w)/P_{tw}(T_w)$, a quotient of quantities that are strange and uncalculable at best. $P_{tw}(T_w)$ is the probability, at t, as calculated from the theory T_w *itself*, of the truth of T_w. Lewis and Hall both note this difficulty with NP, though they don't give it as much consideration as it deserves. There would seem to be no reason why a Lewisian best system T_w must give us probabilistic laws so strong and comprehensive that they determine, at every moment of time, their own probability of truth. One can see how a comprehensive theory of chance physical events *could* define this self-probability. It could assign a probability to every future world-history logically compatible with itself and with the history H_{tw} up to time t; the sum of the probabilities of the non-undermining world-histories would then be the desired probability.[8] The conditions needed to guarantee the definedness of this self-probability, however, would seem to be awfully strong. Quantum mechanics, for example, doesn't obviously have the ability to give us probabilities of whole world-histories.

To assume that the Humean chances must cover virtually all expressible propositions, and in particular, cover themselves (as in $P_{tw}(AT_w)/P_{tw}(T_w)$) is to make a strong and unwarranted reductionist assumption. The assumption is partly explained by Lewis' adoption of the "package view" of laws and chance together. For Lewis, in a world like ours, the *real* chances are the ones found in probabilistic fundamental physical laws. By assuming that these laws govern everything that happens (which even BSA-package laws need not do, of course), and enough reductionism to get all macro-probabilities entailed by fundamental micro-probabilities, he comes to see it as plausible that just about everything you can think of has an objective BSA-chance.

---

[8] Here, as is so often the case, we have to assume that (at least some of) the future histories have a finite number of chancy events, so that the terms of this sum are not all zero.

But there can be no rationale for demanding that *every* candidate Best System theory be of this type, and obviously, we have no good reason at all to suppose that our world admits of a such a theory.[9] As we will see in section 6.1 of chapter 6, there are also reasons to doubt the macro-to-micro reductionism needed to make it seem plausible that a fundamental-physics system entails all the objective macro-level chances we need, such as the chance of drawing an ace from a shuffled deck of cards.

So the NP solution cannot be put forward as a general solution even for Lewisian chance; still less can it be assumed to make sense for HOC in the actual world. What do we say, then, about the undermining contradiction problem?

The answer I proposed in (1997) lay in recognizing the inherent limitation of scope of the applicability of Humean objective chances. Humean chances are the kind of facts that, if you know them, are optimally useful for making predictions/bets about events associated with a chance setup, independently of where and when one finds oneself in the world. But as we saw in section 4.1 of chapter 4, it is crucial to the story that we are considering making predictions or bets concerning very small bits of the Humean Mosaic (HM), given the fact that there is a chance-making pattern that exists over the *whole* mosaic. Predictions or bets about huge fragments of history, or sequences of events large enough that they can make or break the chance-patterns at issue, are not, I argued, part of the "intended use," so to speak, of Humean chance. I illustrated this with a Lewis-style toy example:

> *Scenario:* The end of the world is near. There are just 10 more chance events of type Q left in history (think of them as being, in structure,

---

[9] It is worth noting, also, that if the fundamental laws are deterministic—whether they are understood as BSA laws or some other kind—then we have even less reason to expect the chances to cover everything; they may cover only a few types of oft-repeated macro-scale phenomena. Loewer proposes that in a Newtonian particle world suitable for systematization with classical statistical mechanics and thermodynamics, it may be that the Best System postulates a certain static probability distribution over the space of initial conditions (the microcanonical distribution). This might get us the required probabilities-of-histories, though it still requires a very strong reductionist assumption. Whether Loewer's proposal works is discussed in (Frigg 2010).

like flips of a coin). In past history, there have been 30; and exactly 15 were "heads," nicely distributed temporally. Depending on how the last ten turnout, the probabilistic laws may take form T (if anywhere from 1 to 9 Q events are "heads"), T' (if 0 are "heads"), or T" (if 10 are "heads"). For concreteness, let's suppose these laws are as follows:

T: Each Q has chance $\frac{1}{2}$ of being "heads," and $\frac{1}{2}$ of being "tails."

T': Each Q has chance $\frac{1}{3}$ of being "heads," and $\frac{2}{3}$ of being "tails."

T": Each Q has chance $\frac{2}{3}$ of being "heads," and $\frac{1}{3}$ of being "tails."

And: E is the statement that there are 10 more Q-events to go in history. In fact, in the world in question, 4 are "heads" and T are the laws. But T assigns a non-zero chance to both the 0-heads and 10-heads outcomes. T also assigns much higher probability to 4-, 5- or 6-heads outcomes than it does to 1-head or 2-heads outcomes, and so on.

Question: What credences in possible future outcomes should a good Humean, knowing H_{tw}, T_w and E in these circumstances, have?

NP Answer: Distribute credence over the outcome-types from 1-head to 9-heads. Each outcome gets a boost compared to its objective probability under T. The "middle"-outcomes get more boost than the "fringe"-outcomes.

Right Answer: A good Humean knows only that one of the 1-9-"heads"-outcomes will happen. She has *no reason whatsoever* to assign more credence to one than to another. (p. 329)

Why not? She knows that she is pondering a set of events of such importance (*vis à vis* the HM) that it helps determine what the actual chances turn out to be. But the chance law which we are assuming she knows is true is the *result* of the HM, not the *producer* of it. Our Humean has to mentally "step back" a bit, and remind herself that when it comes to these 10 remaining events, *che será será*, and that is all that can be said. For *these* events there seems to be no rational reason to prefer one way of assigning credences over another, as long as consistency with the known law is respected. The situation is quite different, of course, in the standard case of considering what credences to have regarding a *small* bit of the HM, with no potential for undermining at all, and about which one knows nothing other than that it is part of an overall chance-making pattern, and that the chance of A is x. In this sort of situation it is possible to show that the best credence function to have is one that assigns credence x to A—as we saw when we deduced the validity of PP. What the toy example appears to teach us is that the domain of validity of PP—or, equivalently, the domain of applicability of Humean chance—is intrinsically limited. The restriction to small parts of the overall pattern of events is crucial; it is part of the very *concept* of Humean objective chance, I argued, because it is what allows the justification of PP, which itself is the core of our concept of chance.

And with this, I claimed, the solution to the undermining contradiction was clear. The solution to the undermining problem is to ignore it: it can only arise for situations to which Humean objective chances were never meant to apply.

♥

Unfortunately, I now think that this solution is not tenable after all, and much more needs to be said to truly solve the undermining problem for HOC. A different resolution of the problem will be offered in section 5.5. But to lay the groundwork for the solution, I want to first go back and look a bit harder at the way undermining arises. In all the examples I know of, the key to the creation of undermining sequences lies in the *independence* of the individual chance events that make up the sequences. But what, for a Humean, is this notion of independence? We will begin to approach the issue by thinking about independence in the context of a different Humean theory of probability, actual frequentism.

5.4. Independence for Humeans

Typical examples of undermining are constructed by concatenating long strings of events falling under a chance rule—coin flips, radium decays, etc. But from knowing that $\Pr(\text{H}|\text{Flip}) = 0.5$, how do we infer $\Pr(10^3 \text{ Heads-in-a-row}|10^3 \text{ Flips})$? We posit the *independence* of the flips, which gives us (for example) Pr(H on flip 3|H on 1&2) = 0.5, and so forth for any string of prior results we care to put on the right side of the conditionalization.[10]

But when you stop to think about it, for an actual frequentist account of objective probabilities, independence is a strange thing to posit. Strange, or simply incoherent. The whole point of actual frequentism is to insist that the probability, for example, of heads on flipping a fair coin is just the total number of heads results in the class of all such flips, divided by the total number of flips. So the probability of heads on this next flip, far from being independent of the outcomes of other flips, is in a sense a mathematical function of those outcomes![11] This makes undermining cases even easier to produce for actual frequentism than for HOC. Conversely, given the conceptual affinities between actual frequentism and HOC, it shows us that independence is a notion that must be treated with care in HOC.

Either an actual frequentist, or a proponent of HOC, can simply insist that individual trials of a chance setup S *are* independent. The frequentist can point out that the actual frequency is simply a numerical fact that determines what the objective probability is; it's not as though the results of other trials reach out over space and time to exert an invisible influence on things like coin flips (exerting, on the last flip in history, an iron control forcing the flip to have the right outcome to produce the correct total frequency). And the HOC proponent can say something similar, with the additional point that in general, actual frequencies do not have to agree perfectly with the HOCs.

Nevertheless there is something puzzling here, and undermining gets at the heart of it. Undermining occurs, however, on the credence side of things rather than the chance side. It's not that the actual frequentist ought to (but does not) concede that, on the last flip of

---

[10] It is not *necessary* to postulate independence; the axioms of probability allow one to calculate the probability of a long string of events that are non-independent too, as long as all the joint probabilities required are defined. But independence is the usual assumption made.

[11] Plus the outcome of the flip in question itself. This "non-locality" of frequentist chances is sometimes presented as a serious argument against actual frequentism, e.g., in (Hájek, 1996). Belot's (2016) argument can be seen as the result of taking this observation too seriously.

a coin, Pr(H|all-prior-flip-results) = 0 or 1 instead of 0.5. The actual frequentist can postulate independence for all trials, and justly insist that Pr(H|all-prior-flip-results) = 0.5, full stop. No, the problem is that $Cr(H|ch(H)=\frac{1}{2}$ & all-but-the-last-flip-results) has to be 0 or 1, by the rules of logic and the subjective probability axioms. And this, together with the assumption that PP holds for the actual frequentist's chances, is what generates the undermining contradiction.[12] In the case of HOC, one cannot get undermining contradictions quite so easily and ubiquitously, but they exist all the same.

One might think that the problem lies in the overreaching supposition of independence for arbitrarily large sequences of Humean chance events. Rather than postulating independence across the board, what if we restricted it to collections or sequences of length N or less—where (for the HOC chances in S, at least), no possible outcome sequence of length N or less has the potential to undermine the chances? This is in effect what I proposed in my 1997 discussion. HOC should be considered applicable to small chunks of the outcome pattern, but not to chunks large enough to potentially house outcome sequences that undermine the chances. What about those chunks of outcomes of length N or greater? How should our credences be distributed over them (aside from setting credence in undermining outcomes to zero, of course)? My claim was that "anything goes"—we can assign credences any way we like. Since the Humean recognizes that the pattern determines the chances and not vice versa, there is no reason to assign chances over the non-underminers in a way dictated by PP (or rather, NP, since it incorporates the needed "boosts" to get all the probabilities adding up to 1).

I now see that this was a mistake, in particular the claim that we have no reason to distribute credences over the non-undermining outcomes one way rather than another, when dealing with >N outcome

---

[12] I realize that objections can be made to this sort of undermining case, along the lines of Loewer's and/or Roberts' discussions. How could a real epistemic agent know the total number of coin flips in universal history? How could an agent know the true objective chance (= actual frequency)? But set aside these concerns about the specific examples of undermining; they are not really relevant as long as we are sure that undermining cases *can* be constructed that involve no cheating. And our example of 90% heads outcomes in the next millennium of human history gave us such a case.

sequences. This *is* a plausible thing to say, in the specific case of my toy example—where we are dealing with only 10 outcomes, in a pattern that as a whole contains very few more. But in more typical cases like coin flips and dice rolls in the *actual* world, the "anything goes" claim is quite mistaken. It would entail, for example, that a Humean could assign exceptionally high credence to one *particular* >N sequence of outcomes, assigning near-zero credence in all the others—as long as that sequence is not an underminer. But that is crazy, and inconsistent with what we need to be able to say about small-ish subsequences of results *within* that >N sequence: that HOC and PP apply perfectly well to them, and that our credences in the possible outcomes there (all non-underminers) should be precisely equal to the objective chances.

"Anything goes" is not on. Rather, it seems clear that we want to assign credences over the possible (non-undermining) outcomes in the >N sequence of trials in a way that is as close to PP's prescriptions as possible, but boosting credence in these outcomes slightly to compensate the zeroing out of the underminers. But this is precisely what the Lewis-Hall NP solution does! Perhaps they were correct after all.

5.5. The NP Solution Redeemed

Earlier I criticized the NP solution for demanding that theories of chance be strong enough to have a huge domain—in particular, a domain that includes the theory itself. Recall *NP*: $C(A|H_{tw}T_w) = x = Pr(A|T_w)$. T_w represents the *entire* theory of chance for world w. Demanding that the probability of T_w exist is bad enough (why should it?), but still worse is the fact that, in any world that is complex like ours, NP tells you to set your credences to a number that only a Laplacean demon could calculate—if that number exists at all.

If you look at the toy examples of undermining usually discussed (e.g., in Lewis 1994), and how NP is applied to them, you will notice that $Pr(T_w)$ is calculated by assuming that it corresponds to the sum of the probabilities of all the non-undermining outcomes that complete the chance-making pattern for the world. In the postulated toy worlds, that's all the future there is. But if the world is more complicated, and T_w is a theory that covers more than just the one type of chance at issue, then $Pr(T_w)$ will *not* be equal simply to the sum of the non-undermining futures of the chance setup at issue! It will be more complicated—if it exists at all.[13]

So are the toy-example calculations of NP wrong, or applicable only in bare universes with only one type of chance process? In my view, it is not the calculations that are mistaken, but rather NP itself. NP tells you to conditionalize Pr on the truth of T_w, but what you should do is actually something much less ambitious: conditionalize on the non-occurrence of an undermining result, in the event or events over which credence is being assigned by PP. So if a sequence of N trials of setup S is at issue, and some possible outcome sequences are underminers, one should zero them out and redistribute credence; no less, but also no *more* than that. That is what actually is done in Lewis' (1994) toy example.

So what I view as the proper solution to the undermining problem is this. Let $\{A_i\}$ be the set of possible outcomes we are contemplating, among which the subset $\{A_u\}$ are outcomes that, in conjunction with E, are underminers of the chance rule X. Then for all A_i,T

$$NP^*: \quad C(A_i \mid XE) = \Pr_X(A_i \mid \neg\{A_u\})$$

And the correct analog of GPP, for agents uncertain about the correct chance rule for A, is:

$$GNP^*: \quad C(A \mid E) = \sum_i C(X_i \mid E) \cdot \Pr_{Xi}(A \mid \neg\{A_u\})$$

In words, NP* says: set credence in any possible outcome of a chance setup equal to the objective chance of that outcome *conditional on the non-occurrence of an undermining outcome*. What determines what counts as an undermining outcome? The answer is simple: if the conjunction of A_i, X and E is a contradiction (given that the meaning

---

[13] As remarked earlier, there is no reason to think that T_w must be complete enough to dictate, at every moment of time, its own chance of turning out to be the correct chance theory. It is also instructive to note that in giving his toy example of NP at work, Lewis (1994) assumes finite frequentism rather than his own HOC. Thereby he avoids having to discuss the thorny question of how a chance theory, in general, assigns its own truth an objective probability.

of "chance" is given by the Best Systems analysis), then A_i belongs in $\{A_u\}$; if not, not.

NP* has just the features we want. It solves the undermining contradiction problem: any F-type outcome will get zero credence, since $Pr(F|\neg F) = 0$. Non-undermining outcomes, in scenarios where *no* possible outcome is an underminer, get credence equal to the chance: since $\{A_u\}$ is empty in such cases, NP* reduces to PP. And finally, where undermining does loom as a possibility, the non-undermining outcomes are assigned credence equal to the chance, conditioned on the failure of undermining to occur. Again, this is just what Lewis' 1994 toy example does.[14]

What about when undermining does not directly loom, but one contemplates small sequences of events that *could* form part of a larger future sequence of events that could be an underminer? Shouldn't credence in such sequences be tweaked slightly, to take account of this fact? My 1997 answer was *No*, and I still believe that was right. Consider my toy scenario. There are just 10 "flips" left to go, but now we want to assign credence to how the next five come out. Surely we should not just use PP (or NP*) and assign probability $\frac{1}{32}$ to an outcome of 5 heads in a row—since that is half-way to being an undermining sequence of 10 flips, and hence (for that reason) "less likely" to come about? I think this argument can be resisted. One might worry that applying PP simpliciter to such 5-toss sequences will quickly land us in trouble. For if the probability of the next 5 flips giving heads is $\frac{1}{32}$, and the probability of the last five flips giving heads is $\frac{1}{32}$, doesn't that mean that the probability of all 10 flips giving heads is $\frac{1}{1024}$—given independence?

On the probability side, yes; and that is just fine. But credence is not objective probability, and it need not display the same independencies as probabilities do; indeed, given undermining, it cannot do so! So while a rational agent may assign: $Cr(\text{next-5-H}|XE) = \frac{1}{32}$, and

[14] In order to do his toy calculation with NP, Lewis (1994) had to assume that we know *precisely* how many future *S-setup* events there will be. This was not a mere simplifying assumption; without it, it is unclear how one would begin to do the calculation, even for a world with just one chance law. Lewis' toy chance rule itself makes no predictions about *how many future chance events will exist*, and without this information it is unable to assign any particular chance to its own undermining or non-undermining.

$Cr(\text{last-5-H}|XE) = \frac{1}{32}$, both in accordance with PP and NP*, this does not entail that for the agent, $Cr(\text{last-5-H}|\text{next-5-H \&XE}) = \frac{1}{1024}$. This last line violates NP*. Adding "next-5-H" to X and E has the effect of making last-5-H a sure underminer of X; so NP* says this conditional credence should be zero.

NP* in essence is a rule about credence that, I think, any Humean must accept, given that undermining is a genuine possibility. It faces up to the fact that some specifiable events or outcomes are jointly inconsistent with the Humean chances being X and the rest of the agent's knowledge E; in such cases, credence must be zeroed out, and so credence in alternative outcomes that do not have this inconsistency problem must be boosted to compensate. But whenever the event A whose chance one is interested in is something small compared to the overall chance-making pattern, which is usually the case, there are *no* underminers in the space of possible outcomes at issue, and NP* reduces to PP.

Which brings us back to at least part of my (1997) claims: the correct solution to the problem of undermining is to ignore it! Not at the conceptual level, of course; but at the practical level, in a big world such as ours, NP* almost always reduces to PP, and where it does not, the corrections it introduces are normally going to be miniscule.

5.6. Summing Up

We have seen in this chapter that the correct solution involves modifying the credences dictated by the chance system, wherever the potential for undermining futures exists. Hall and Lewis' NP solution was on the right track, but from my perspective went too far in urging that rational agents conditionalize on the truth of the entire chance Best System, when in fact all that is needed is to conditionalize on the non-occurrence of undermining *for the chance rule X*.

♦

Box 5.1 Does NP* Need a Separate Justification?

In chapter 4 we saw that the original PP could be justified by two sorts of argument, one "consequentialist" in nature and the other *a priori*. But now we are abandoning PP because of undermining and replacing it with NP*. If PP is rationally required, how can NP* even be tolerated, much less rationally required, given that in some cases at least they conflict? And even if the conflict can be massaged away somehow, don't we need a new chapter showing (if possible!) that rational agents must follow NP*?

No need to panic. Looking back at the arguments for PP, we recall that both have features that, in effect, put limitations on the scope of what was established. In both, what was established was the rational requiredness of conforming to PP for ordinary small-scope applications of chance rules covering easily repeatable setups—things like dice rolls, roulette spins, radioactive decays, and so forth. In other words, the arguments showed the rationality of PP for precisely the kinds of applications where NP* collapses into PP, and hence they supported NP* just as much as PP. But let's look into this further and see if we can answer the question of which principle is justifiable for the *unusual, large-scale* events where undermining becomes a possibility.

Consider first the consequentialist argument that I adapted from Strevens (1999). We considered an agent who has to place multiple bets on outcomes in medium-sized runs of trials of the chance process, such as placing bets on the outcomes of, say, 50 consecutive coin flips. Because of the way HOC is guaranteed to keep the objective chances near to the frequencies in randomly chosen collections of such medium-sized runs, we could see that the agent has no reliably *better* strategy, if betting using some fixed odds, than making the chosen odds equal to the chances.[a] But it is clear that this sort of scenario is one in which undermining is not on the table, and so NP* coincides with PP. Runs of 50 heads in a row are unusual, indeed possibly non-existent in human history, but they are no underminers; so nothing being contemplated in the scenario raises any question of needing to zero out credence in some outcomes. (In fact, the right way to think of the scenario is actually as one in which the agent bets on a *single* coin toss, conforming (or failing to conform) to PP on each toss, but does so 50 times. And it is as clear as can be that for a *single* coin toss NP* does not differ from PP.)

The *a priori* argument for PP considered, instead, what an agent can coherently *think* are fair odds for wagering on a repeatable chance setup, if she knows the Humean objective chance. The conclusion of that argument was that the agent has incoherent—logically inconsistent—beliefs if her credence level y (and hence, what she judges to be fair odds for bettors on either side) is different from what she believes the chance to be. The reason for this is that to have credence in outcome A in S corresponding to odds of $y{:}1{-}y$ entails judging that in an indefinitely long sequence of wagers on $A/\neg A$ at those odds, there is no advantage to one side or another (given our understanding of credence and the law of large numbers). But understanding the theory of HOC, an agent knows that in an indefinitely long sequence of trials, the frequency of A must be extremely close to x, the objective chance of A.

Now, an implicit assumption here is that A is an ordinary small-scale event of the type that we expect our Best System to clearly and directly assign chances (e.g., coin flips or rolls of a die), easily repeatable events. For such things, there is no outcome that is a potential underminer, so NP* collapses into PP and the *a priori* argument justifies both principles equally.

To see whether the *a priori* argument justifies NP* more generally, we need to consider outcomes F that are of a "size" (relative to the underlying chance rule used to calculate the chance) where undermining looms as a possibility. We might let F be the outcome: $<\frac{1}{2}$ of all radon atoms on Earth decay in the next hour> (this being an extraordinarily improbable event, and one which (we will assume) would, if actual, force the Best System to have a special chance rule governing radon-decays on Earth from such-and-such time onward). Now, a moment's thought shows us that our rational agent must give F zero credence (conditional on history to date plus the truth of the actual radon decay chance rule), and hence boost her credence in some or all of the propositions specifying *non*-undermining behaviors of Earth's radon atoms in the next hour. So even a non-undermining outcome such as *A:* <0.001% fewer radon atoms decay on Earth in the next hour than one would expect based on the quantum chance> may be such that the rational agent must assign it credence which violates—infinitesimally!—the original PP.

I say "may be such" here because at this point all we can see is that the agent *cannot* follow PP strictly and must instead have zero credence in any underminers, such as *F*, and redistribute credence among the non-undermining possible outcomes somehow or other. NP* is *one* way to redistribute credence, but we have not yet shown that it is rationally required. Still, by considering the *a priori* argument for PP from chapter 4 with an undermining chance event *F*, we have shown that there is a loophole in (or limitation of) that argument that we did not notice: it does not demonstrate the validity of PP *tout court*, but only in the context of credences about chancy events of a size/scope such that undermining is not an issue. And we have seen this simply by considering the first stage of the *a priori* argument, so to speak: just thinking about the agent's credences in the possible outcomes of *one* instantiation of the setup *S*, never mind what happens if *S* is instantiated again and again in an indefinitely long series.

In order to see whether the *a priori* argument validates NP* or not, let's now consider precisely an indefinitely long series of instantiations of *S*, the letting-evolve of all the radon atoms on Earth for one hour. Now, what NP* counsels is that the agent have credences equal to the objective chances *conditional on* $\neg\{A_u\}$. Let's imagine that the agent's credences do not follow NP*, but instead differ from that prescription. Can we descry a contradiction here?

That depends on how significant the deviation from NP* is. As we saw in the *a priori* argument for PP under HOC, there is a bit of "wiggle room" for the agent: she is only incoherent if her credences mismatch *non-trivially* what she takes to be the chances, because she knows that if a non-trivial frequency mismatch were to be found in actual outcomes over an *indefinitely long* sequence of trials, the nature of the Best System analysis would entail that the objective chance rule for *S* be different than as hypothesized. (How big *non-trivial* is may vary from setup type to setup type, and in any event it is probably not sharply definable.) But in general, the wiggle room for credence left open by the *a priori* argument is going to be, intuitively, bigger than the difference between the prescriptions of NP* and PP to begin with, or than the difference between the prescriptions of NP* and any other prescription that simply starts from PP's prescription and redistributes the credence from undermining events over the remaining events in whatever fashion you like.

What we can say, then, is this: to the extent—limited, but plenty good enough—that the *a priori* argument showed PP to be rationally required, it also *thereby* showed NP* to be rationally required. This is so for any real-world situation where we can imagine an undermining outcome as possible and therefore find the counsels of NP* to be (ever-so-slightly) different from those of PP. The only setting in which there can be a non-trivial difference between PP's prescriptions and NP*'s prescriptions is in the case of toy examples like that of section 5.3, where the agent somehow knows that the chance-making pattern has almost come to an end. But in such a situation, the agent's credence is prevented from entailing anything about the fair betting odds in an indefinitely long sequence of trials—there will be no such sequence—so the *a priori* argument did not really cover such cases.

When it comes to credence in underminers *F* in general, NP* has the right of it, not PP; so we can say: where NP* differs from PP, it is NP* that is correct. But in non-toy example situations, NP*-credences differ from those of PP by, at most, a trivial amount, less than the "wiggle room" we noted that the *a priori* argument inevitably left open. Therefore the *a priori* argument was—all along, though we did not notice it!—just as much an argument for the rational requiredness of conforming to NP*.

---

[a] Or, if the opponent sets the odds, the agent will do best to choose to take the side of the bet which PP recommends, whenever the opponent sets odds that do not correspond to the objective chances.

6

Macro-Level and
Micro-Level Chances

The undermining problem we considered in chapter 5 puts, in a sense, limits on the scope of applicability of Humean chances via the standard PP: if one is considering the probability of some large-scale, exotic event "unlikely" enough to constitute an underminer for a chance rule that one believes (or assigns non-zero credence to), then the credence assigned to that event (conditional on the chance rule, and the standard setup of the rule) should not be x as PP says, but instead *zero*. This is a "limitation" in a theoretical sense, but not one we will ever need to worry about in real practice.

But thinking about the oddities that can result from taking basic chance rules and applying them to huge Boolean constructs led me to consider the question of whether there might not be *other* perils lurking in such remote applications, perils different from the undermining contradiction problem. This chapter concerns such a further peril—albeit again only a theoretical peril, not one we will ever bump into in practice. It is a look at the problematic nature of chances for macroscopic events derived from micro-level chances.

6.1. A Chance for All Setups?

From the apparent existence of micro-level objective chances, many philosophers would like to infer the existence of macro-level chances *entailed by* these micro-chances, and claim that these entailed chances are the "true" objective probabilities. This seems to have been David Lewis' belief, and many metaphysicians who postulate objective chance appear to share this chance-reductionism. In this chapter, we will see why it is misguided in several ways.

The reductionist/supervenientists believe that if there is a certain sort of completeness to the micro-level chances, then they will automatically entail objective chances for just about all propositions specifying outcomes, given a well-specified setup or initial state of affairs.[1] The sort of completeness needed is something like this: (a) there are complete states of the world-at-a-time (or big-enough regions of the world) that fall under the scope of the chance micro-laws; (b) given any such initial configuration,[2] these laws entail an exact probability for any specifiable future configuration to evolve from the initial state; (c) a Laplacean demon could in principle decide which of these possible-future-configurations, or which regions of phase space, correspond to states of affairs in which the proposition in question is true. The Laplacean demon just has to add up or integrate over the appropriate possible states using the micro-laws, to obtain the *true* probability that the proposition shall obtain. The details of how these ideas are implemented may be important, but for now let's not try to fill them in. Instead let's ask: what's wrong with this picture?

---

[1] Lewis clearly had something like this macro-micro reductionist approach to chance in mind. It is also, in a slightly different form, advocated by Loewer and Albert in their proposal to derive probabilities in statistical mechanics from the imposition of a probability distribution over initial conditions of the universe compatible with a (specific sort of) low-entropy macro-state. See (Albert, 2015), and Loewer's chapter in (Barry Loewer, Weslake, & Winsberg, n.d.).

[2] In Albert and Loewer's approach, the initial configuration is actually an initial macro-state posited at the beginning of time, plus the standard Lebesgue measure restricted to that macro-state in phase space. The measure then allows the Laplacean demon to calculate a probability for evolution into any sort of future macro-state that one cares to specify. The fact that their way of deriving chances from fundamental physics goes via statistical mechanics and macro-states does not make the required calculations any easier; in fact, the contrary is the case.

In the deterministic classical physics they are assuming, there is also a fact about the universe's precise initial micro-state—a point in phase space—but that plays no role in determining the probabilities.

6.2. Epistemology of Micro-Derived Chances

First and most obviously, there is its irrelevance to human concerns. We are not Laplacean demons, and we will never, *ever* be able even to estimate a probability such as, for example, the probability of *my* contracting lung cancer within the next five years, from microphysical laws and initial conditions. One might at first think that it involves only something *extremely difficult*: knowing the current state of my body in complete micro-physical detail, and calculating the probability of lung-cancer-causing mutations on the basis of that current micro-state. But a moment's reflection reveals that things are vastly more complicated than that. My probability of developing a lung cancer depends, obviously, on what I will breathe in during the coming years. And that depends on a myriad of things: where I will travel, what routes I will take as I walk around, whether I take up running or not, etc. It also depends on what x-rays and cosmic rays will be intercepted by my body, what radon atoms breathed in, and so forth. In fact, at a *minimum*, the initial conditions required in order to calculate the desired probability from the micro-physical initial state would be the *complete* state of the world around me in a five-light-year radius![3]

This epistemic point is so obvious that it should be unworthy of mention. But surprisingly often, philosophers act as though they think they can make good estimates of the demon's calculations, at least in a comparative sense. Lewis is a good example: he bases a theory of causality on facts concerning whether the presence or absence of a certain factor *C* raises or lowers the probability of a posterior event *E*. And Lewis accepts the micro-chance-supervenience picture wholeheartedly. So he obviously believes that, while we can't calculate the precise probability of *E* in a world just like ours in which *C* is removed, we can be *sure*, in many cases, that it is lower. On what grounds?

---

[3] I say "at least" because it is an open possibility that, due to quantum entanglement, states of affairs even further away than five light-years might be causally (or at least probabilistically) relevant to what events will occur inside the five-light-year-radius sphere.

To this question, the reductionist philosopher will perhaps react with impatience. Surely it is obvious, in normal cases of causation, that the presence of C elevates the probability of E occurring, over what it would be were C absent? Well, it *seems* obvious—because we already think of C as a cause of E, and antecedently think of introducing a cause as an effective strategy for bringing about the effect. In effect, we move from believing that C is an effective strategy for E, to the natural conclusion that C must raise E's probability.[4] But this is in reality no inferential move at all: in the context of a (presumably) not-total, non-deterministic cause C, being an effective strategy for E *just is* being a probability-raiser of E in the context. In short, we have no grounds here to support the claim that C raises the probability of E, only a re-statement of the belief that it does.[5]

And we have no other, genuine evidence to which we might appeal, to justify that the probability of E with C is higher than without C. The plausibility of the probability-increase cannot be defended on inductive grounds, by looking at (macroscopically) similar cases. If the micro-derived probability in question is single-instance—i.e., the setup has only one instance in all history—then this is obvious: there is only one case to look at, so not enough to make any sort of induction. But even if the micro-derived chance were somehow multiply instantiated, its value could be estimated from the actual frequency only if the PP is correctly applicable to these chances. We will see in the following that this is very doubtful.

6.3. Micro-Derived Chances and PP

Throughout this chapter we will be assuming that chances for macro-level events are derivable *in principle* (by a Laplacean demon) from

---

[4] We will see how common and natural such inferences are, in chapter 8.

[5] It is a familiar point from the counterfactual-causation literature that whether the probability of C with C is higher or not may depend on very precise details concerning the manner in which C is counterfactually deleted. The details at issue are macroscopic. But it seems to me likely that microscopic details could easily make all the difference between E's occurring or not, also. I argue for this in (C. Hoefer, 2004).

micro-level chances, i.e., we will be assuming a fairly strong reductionist supervenience. Serious philosophical objections can of course be raised against this reductionist claim, as is done for example in (Dupré, 1993). But setting aside the metaphysical debate, I want now to discuss a second problem with micro-derived chances of macro-level events, a problem that we might call "epistemic." But here, the epistemic problem at issue will not be our inability to know or derive such chances, but rather the question of whether such chances have epistemic normative force: can they legitimately serve as belief-guides in the way captured by the Principal Principle (PP)? I believe the answer is, in general, negative. When we see why this is so, we will see that there are further limitations on the scope of applicability of Humean chance, limitations not directly related to undermining, but rather to the extension-by-calculation of chances to realms far beyond their native home.

Let's discuss two different sorts of micro-derived chances in turn: very specific, single-case chances, and more generic chances. Single-case chances are the sort of history-dependent, localized chances that both Lewis (1994) and Mellor (1995) believe we should countenance. Here is an example.

Q: "Carl drinks two or more beers on the evening of April 23, 2017"
X: "$Pr(Q|$state-of-world at $[t_0 = 12:00$ a.m., April 22, 2017$]) = q$"

By contrast, micro-derived *generic* chances—e.g., $Pr(H|$Fairly flipped coin$)$—present certain complications and quandaries; we will come back to them later.

Suppose that we grant for the sake of argument that single-case chances such as X are indeed entailed by the micro-chances in the Best System. Should such chances guide our credences, if Laplace's demon whispered them in our ears? I believe it is hard to maintain that they should. We can see why, by recalling our two justifications of PP from chapter 4.

The pragmatic argument turned on the fact that most chance setups will be instantiated a large number of times in world history, with the outcomes distributed in a nicely "chancy-looking" way over space and time. But for any proposition like Q concerning a one-off event, there can be no multiple trials, and the actual frequency will be 1 or 0.[6] All single-case micro-derived chances share this feature, so the pragmatic justification of PP simply fails to apply directly. Might it apply in a less direct fashion?

It might, in the following way. Suppose we group together all the specifiable outcomes P_k that are, or might be, of interest to humans, and that at t_0 all have micro-derived chance of 0.15 (±.0005). (The index serves just to number and distinguish all these various propositions stating one-off events.) There would presumably be lots of these in a complex world such as ours. Then, if the frequency of occurrence of the outcome in this class is very close to 0.15, and the pattern of occurrences is stably distributed in the normal chancy way, one might then say that the pragmatic argument should function here just as well as it did in the kinds of case we studied in chapter 5. If the same goes for each probability $0.x$ (spaced, say, in steps of .0005), then *voilà*: the pragmatic argument can serve to justify PP for micro-derived chances of one-off propositions concerning outcomes of interest to humans.

This *would* be the case, *if* the frequencies did in fact line up with the chances in this way; but do they? We will never know! Each one of these micro-derived chances is beyond our power to calculate, or even estimate, so we will simply never know how well the frequencies match the chances, if they do at all. In the following we will see how they might *not* match the chances, without this undermining the Best System of chances for our world in the least.

What about the a priori argument for PP under HOC? It also is in-conclusive for micro-derived chances. The point of the a priori argument can be re-expressed this way. Can we coherently imagine these two things being true together: that the HOC $Pr(A|S) = x$, **and** that in an indefinitely long (but finite) sequence of trials of S, $Freq(A|S) = y \neq x$ (where y is non-trivially different from x)? When S is a setup that figures directly in the Best System—i.e., it has its own axiom, so to

---

[6] I set aside here the intriguing (maddening?) thought that if our universe is spatially infinite, perhaps we should expect any specific physical condition (or something rele-vantly near-identical) to occur at many places (/times) in the universe. I believe that this does not in fact follow, even if we grant the completeness of a chancy micro-physics. In order to facilitate setting aside such flights of speculation, let us assume that our uni-verse is in fact spatially finite, although very, very large.

speak—the answer is *No*, we can't coherently conceive both of these being true, because of the nature of HOC with its intrinsic limitations on frequency tolerance. But now we are considering micro-derived chances that do not figure directly in the Best System; in fact they are vastly far removed from the axioms of the system (presumably chance "laws" something like what quantum mechanics gives us). Micro-deriving chances for macro-scale events requires calculations of fearsome complexity, hugely far removed from the simpler applications of quantum physics that are used to test the theory's predictions.

Now if *S* is a hugely complex physical setup (e.g., the state of everything within a five-light-day radius of the Earth, *now*), and *Q* is a macro-sized outcome that can be subvened by a huge number of possible micro-specified states of affairs posterior to *now*, we are talking about something quite abstruse and far removed from the more "typical" and repeatable states of affairs whose patterns make it the case that our hypothesized QM-like rules are part of the Best System's chances (things like individual radium decays, absorption and re-radiation of a photon by a helium atom, etc.). So let's ask our hypothetical question from the a priori argument for PP again. Can we coherently imagine these two things being true together: that the HOC $Pr(Q|S) = x$, **and** that in an indefinitely long (but finite) sequence of trials of *S*, $Freq(Q|S) = y \neq x$? This is not something that we can easily imagine to be instantiated multiple times in our world, of course; the exact state of affairs here/now over a five-light-day radius will never be replicated anywhere in our universe.[7] But never mind that, since it does not affect the conceptual question. *If* we imagine it to somehow be repeated over and over, in a world with the same Humean chances as ours, can we imagine the frequency of Q-outcomes to be $y \neq x$?

---

[7] Notice that inside this five-light-day region of space there is an abundance of what we normally think of as *records* or *evidence* concerning states of affairs farther away/at earlier times: all the information-bearing light rays that are to be found in the (mostly empty) space, which have traveled from far-away stars and galaxies. To imagine the full micro-state being found somewhere else in the universe, one must imagine either (a) the entire history of the world (backward light cone region) is also identical to ours, or close enough; or (b) some bizarre accident makes all the light rays inside the identical twin region *misleadingly* suggest a past history just like ours, even though in fact things are quite different in that region's backward light cone; or (c) some mix of (a) and (b).

It seems to me that the answer is, clearly, *Yes.* The reason is that, while HOC has intrinsically limited frequency tolerance for the chance rules that figure directly/basically in the system, the frequency tolerance of calculated/derived chances for enormously complex events (or enormous complexes of events) is much greater.

To understand why, we need to understand the relationship between X and the underlying micro-event chances from which it is derived (by our hypothesized Laplacean demon). Even a "single-case" chance such as that for the truth of Q amounts to a chance, not of a certain specific micro-state coming to be in a region of spacetime, but rather a huge disjunction of different micro-states, in spacetime regions of varying sizes. The disjunction, of course, is the hypothesized disjunction of micro-states adequate to subvene the state of affairs Q. From the *macroscopic* perspective, all these micro-states have something in common (I drink two or more beers on a certain evening). But from the microscopic perspective, they do not. The number of atoms and other particles, number of each type of atom, specific spatiotemporal arrangement, energy levels of each particle—all of these can vary wildly from disjunct to disjunct, and over the whole set of disjuncts they will indeed vary tremendously.[8] So, vis à vis microscopic states and configurations (the kinds of things for which our micro-chance theory dictates probabilities), the huge disjunction that subvenes Q is an utterly heterogeneous, hodge-podge disjunction of huge, many-particle states—a "gerrymandered" disjunction, we might say.

Getting back to the a priori argument for PP, then, the question we have to consider becomes: can we coherently imagine that the X setup S is repeated indefinitely many times in a world with our Humean chances, yet yields a frequency of Q-outcomes significantly different from x? Clearly, we can coherently imagine this: it would not undermine the overall fit of the Best System by more than an infinitesimal amount, nowhere near enough to force the Best System

---

[8] The particles we're talking about here are not just those of my body, or my body + immediate vicinity; rather they are all the particles in that (say) one-light-day radius of space, the region chosen as sufficient to include all physical processes that might interact with me (and hence affect the probability of occurrence of one or more of the disjuncts adequate to subvene Q). So the overall set of disjuncts we are invoking here is unimaginably large.

to include a special Q-clause telling the Laplace's demon not to calculate the chance of Q from underlying micro-chances and instead to substitute the actual frequency. Adding such a special clause, for a bizarre and unimportant macro-event type, would amount to a significant loss of simplicity, in return for an infinitesimal improvement in fit.

In order to make the point about fit clearer, another toy example will be helpful. Let's model the underlying micro-chancy events with coin flips, so that we can have an easy to state and understand micro-chance theory: T_m: every fairly flipped coin has chance 1/2 of landing heads, and 1/2 of landing tails. Our model "world" is made up of myriads of these coin flips, at different places, but ordered (and dated) in time. We will assume that the total number of actual flips (hence, the total Humean mosaic at the micro-level) is very very large, though not infinite. Now, what can our analog of Q be? I suggest the following: imagine an enormous disjunction of very specific propositions about coin flips, and how they come up. One such disjunct might begin "500 consecutive flips land as follows: HHTHTT. . . ." X's analog would be an entailed probability of this enormous disjunction, conditional on a certain starting-state, i.e., some condition on the history of flips prior to the 500-or-so we are interested in. (Since our coin flips are independent, the conditioning proposition is irrelevant to the probability of the outcome-disjunction, according to T_m, unlike what is the case with the real-world examples Q and X.) These disjunct propositions go on for quite a while, specifying a large number of posterior outcomes, each disjunct such as to be calculable on the basis of T_m. The outcomes specified differ greatly from disjunct to disjunct. This enormous disjunction, we will stipulate, has enough disjuncts to have an overall chance of x (let's say $x = 0.66$), the same as the micro-derived chance of Q in our world. Now our question becomes: is there any reason why a rational person should or must adjust their credence in this disjunction to be precisely 0.66? Again, I believe the answer is No.

The total pattern or sequence of coin flips in our toy world is huge, and overall supports the simple micro-chance theory T_m quite nicely. But we know that for any such individual finite sequence, it is possible to specify a gerrymandered type of subsequence, or a disjunction of such subsequences, whose frequency within the overall pattern is quite different from its T_m-derived chance.[9][10] Where this is so, you would be better off adjusting your credences to match the actual frequency. So, the derived probabilities of large, gerrymandered outcome event types (and a fortiori, enormous disjunctions of them) do not possess any automatic right to guide our credences. Thus, we might say, the micro-derived chance for our gerrymandered disjunction is not a proper objective chance at all (where "proper" means "appropriate for plugging into PP"). By contrast, more small-scale, typical, and non-gerrymandered applications of objective chance will be unproblematic. The relative frequency of four-heads-in-a-row subsequences, among all four-member in-a-row subsequences, will be around $\frac{1}{16}$, and moreover they will be distributed in a random-looking way throughout the full actual sequence. If they were not, there would be problems for the claim that the actual sequence has as its Best System T_m. But the divergence of the frequency of a huge, gerrymandered disjunct from its "proper chance" prejudices the theory T_m's fit by a minute amount, at most.

The point is that any actual, large-scale (but finite) mosaic or pattern that "fits well" with a micro-chance law *cannot* "fit well," in a frequency sense, with *every* possible specification of gerrymandered disjuncts of

---

[9] This is simply a mathematical fact: a finite outcome-pattern cannot display frequencies that match the chances for boolean combinations of *every* size relative to the pattern. Those who recall the difficulties of attempts to define randomness for finite sequences will recognize the problem. If we were instead working with an infinite chance-making pattern, and it satisfied the von Mises definition of a collective, then the frequency of our subsequence-disjuncts would have to match the T_m-derived chance—in the long run. But this would still imply nothing at all about what will happen reliably in any finite chunk of history.

[10] It may seem that the sort of set of outcomes we are describing, to have probability equal to 0.66, must include about $\frac{2}{3}$ of the possible outcome sequences (of the appropriate size/length) under T_m, and therefore that if its frequency in the overall Humean pattern were significantly different, this would be a seriously improbable defect in the pattern, perhaps enough to constitute an underminer. But recall that, in analogy with Q, we specified that Q-analog is true if, *immediately after a certain specific sequence of outcomes occurs* (analog of the state-of-world at t_o), one of the disjuncts from the set then occurs. The overall frequency of occurrence of members of the set (conditional on nothing) may be nearly 0.66, while the frequency of occurrence *immediately after a certain specific sequence of outcomes occurs* is quite different. If the overall pattern is big enough, this can be so even if that "certain specific sequence" occurs a huge number of times in history: the rest of the pattern "washes out" the statistical anomaly, rendering it clearly a non-underminer.

the micro-level events. So these big gerrymandered disjuncts are not really covered by the a priori justification of PP, nor by the pragmatic justification. In effect, they are left outside the scope of the proper application of objective chance. But now, returning to the real world, just about *every* macro-scale proposition will be, relative to the micro-chances that hold among the quarks and leptons, a big gerrymandered disjunct. So, we just don't have any good reason to think that these Laplacean, micro-derived chances *should* govern our credences, if only we could know them. Their relation to rational credence, if indeed they have one, remains quite opaque to us.

At this point the non-Humean about chance may think she has a reply to offer, to justify a micro-chance-to-credence link. She goes modal and metaphysical in a big way, and asks us to look at the patterns of frequencies over *many* possible worlds. If we look at a huge, possibly infinite, collection of possible worlds in which our coin-flip theory T_m holds as the best system, then in the whole collection the relative frequency of our big disjunction holding will be 0.66.[11] And in a huge, possibly infinite collection of possible worlds identical to the actual world (at the time t_0 at which we are considering the micro-derived chance of Q), and having the same best system micro-chances as our world, the relative frequency of Q obtaining as of April 24 will be 0.66.[12]

There are two responses to this inflationary defense. First: what reason do we have for thinking that the relative frequencies in these huge collections of possible worlds should, in fact, be as specified? This epistemic point will be familiar to the weary defenders of hypothetical frequentism and propensity theories. And second, even if these claims about huge collections of possible worlds are true, what bearing does that have on what *our* credences should be, here in the actual world? None at all, that I can see.[13] Neither Humeans

---

[11] Depending on what specific non-Humean theory our reductionist holds, she may have to add here: ". . . with probability 1." Smart non-Humeans will want to avoid this, for obvious reasons.

[12] The alert reader may notice that this modal response is essentially the same idea discussed in Box 1.3 of chapter 1.

[13] I suppose a reductionist opponent might try to reply to this challenge by claiming that the frequencies in these collections of possible worlds *are* relevant, because we

nor non-Humeans about chance can offer a decent justification for conforming our credences to micro-derived chances like X.

♦

Turning to more generic chances, like the chances of breast cancer discussed earlier, we will see that, again, there is no justification for claiming that our credences should be adjusted to the micro-derived chances, if such chances even exist. First let's look at the reasons for uncertainty as to whether they exist. Consider first our core example, $\Pr(\text{H}|\text{Fairly flipped coin}) = x$.[14] Because the setup condition is atemporal—it does not specify some starting moment—it is unclear how the Laplace demon can get started calculating it. What the demon can do, at every moment and place, is calculate the probability that (within a certain time span, say) a coin will be fairly flipped *and* land heads, given the configuration of all particles at the chosen moment t. It can also calculate the probability that a coin gets flipped. Dividing the former by the latter, perhaps it can be taken as calculating the probability that a flipped coin lands heads—but this probability is linked to a specific time and place or region. We would expect it to fluctuate as time varies, if the Diaconis coin-flipping physics model is at all correct: the nearer to an actual flip t is, the more the demon-derived chance of heads will move toward 0 or 1. (Think of it this way: when I am just pulling the coin out of my pocket, the chance may be 0.49; when I place it heads-up on my thumb, the chance has moved to 0.51; once my thumb is released by my forefinger and is about to strike the coin, the chance may be 0.95.) But the coin flip chance may fluctuate for other reasons as well, and the point is not

are "lost in the multiverse" and do not know *which* possible world we in fact live in. Assuming we are in a randomly chosen world from the collection, . . .

Much mischief can be done by imagining ourselves lost in a space of possible worlds in which we have been placed "randomly"; I refer the interested reader to the Doomsday Argument debates for examples. For our purposes here, I simply want to note that I believe it is perfectly consistent and coherent to grant the mentioned facts about frequencies in large collections of possible worlds and yet not conform one's credences to those values.

[14] Here we refer to the actual coin-flip chance rule of the Best System for our world, not to the toy theory T_m discussed in the preceding.

confined to quasi-deterministic things like coin flips. The chance the demon gives us for the 9:37 train arriving in Castelldefels more than five minutes late may oscillate wildly over the course of any given morning. Somehow, if micro-derived chances are to exist, they must be extracted from these constantly fluctuating, spatially variable single-case chances that the demon's tools allow it to handle. I don't know if an appropriate scheme of averaging is possible, to extract a reasonable generic chance from the micro-chances; if one *is* possible, I don't see why two or more distinct schemes should not be possible, that might give different outputs. And finally, even if just one scheme could emerge as the correct one, I don't see why we should expect that the numbers it spits out will be close to the generic chances HOC gives us by looking at the macro-scale.[15] We will discuss this possibility of a micro/macro mismatch further in the following.

So much for time-independent generic chances. Fortunately for the reductionist, many generic chances that one might expect the HOC Best System to give us will themselves be time- and/or place-variable, and so more apt for being given a reductionist derivation. Take as an example generic chance G: Pr(have or develop breast cancer|female human, living in city C, 2017) = x.

This probability is still not so easily translated into terms that our Laplace's demon can calculate. While Q above specified an already existing, specific person and a specific (though multiply realizable, at the particle level!) property that person might or might not come to have in a certain timespan, G appears to specify an outcome for a "generic human female." *That* does not have any specific microscopic translation, whether at a specific instant of time (e.g., 0:00 of January 1, 2017) or in general. Its borders are too vague. What presumably does have a translation into micro-level events, if Q does, is the probability

---

[15] An exception to this: chaotic and symmetry-bearing chance setups that amplify microscopic variations in initial conditions to differing outcomes can be expected to get micro-derived chances that mirror the symmetries—at least, assuming that SP holds regarding whatever microscopic initial conditions are relevant to single-case outcome probabilities the demon calculates. For things like roulette wheels and lottery-ball machines, even if the microphysics applied by the demon is indeterministic, it presumably will not be completely independent of initial conditions (the way, e.g., radioactive decays are thought to be in quantum mechanics).

of having or developing breast cancer *of each individual person already living in the world* at the beginning of the specified timespan. So the Laplace demon can offer us one of two things. It can calculate each individual's probability of breast cancer, and then give us the average for the whole class. Or it can calculate the probability of each possible *frequency* of breast cancer, in the C population, given the state of the world at the start of 2017. If the pattern of these frequency-chances has the right shape, it may correspond to the pattern that one would calculate starting from a certain individual, generic chance of breast cancer, of the sort specified in G. In that case, one could say that the demon's chance-of-frequency calculations reveal what the "true," micro-derived generic chance is, having the same form as G. Let's assume for the moment that one of these strategies for extracting generic chances from the micro-level is acceptable.

Do these micro-derived chances then deserve to guide credence via PP? Just as before, the answer is not necessarily positive. In fact, our toy example has made the point already. For it was a (huge, gerrymandered) disjunction of conjunctions, the whole being such as to have chance 0.66; we posited that it might well have a quite different frequency in actual history (say, 0.45), without this fact prejudicing T_m's claim to be the correct micro-chance theory at all. T_m specified chances in a time-independent way and thus chances calculated using it are not directly analogous to micro-derived chances starting from an initial physical state at a certain moment/over a certain region. But the moral of the story remains the same: in the actual world's Humean mosaic, the actual frequency of breast cancers may be quite different from the frequency that emerges as most likely from the demon's calculations.

In chapter 3 we saw that macro-level Humean chances may exist for many sorts of reasonably well-defined setups, like our example breast-cancer setups. Such chances are of course always going to be near to, if not identical to, the frequencies in the actual world's history (when those frequencies arise from large numbers of cases). But micro-derived chances might diverge significantly from those frequencies, without this divergence prejudicing the micro-level chance theory significantly at all. (This was the lesson extracted from the T_m story.) If such a conflict were in fact to occur, which chances would be the right ones with which to set our credences? Given the stipulated frequency facts, the answer is obvious: the macro-level chances.

In some circumstances, then, macro dominates micro! This result is worth mulling over a bit. It runs contrary to many philosophers' intuitions about objective probabilities. Sober, for instance, declares it impossible outright, as a violation of the principle of total evidence (Sober, 2010). And you might think that, even if it is *conceivable* that micro-derived chances could fail to be reflected in the frequencies, it is surely an *unlikely and infrequent* occurrence. But both of these ideas are based on non-Humean notions of what objective chances are. In the case of Sober, he treats the macro-level chances and the micro-derived chances as simply two different conditional probabilities, the first being a probability conditional on a macroscopic description of the initial setup, the second being conditional on the exact micro-state of the initial setup (which entails the macro-state).

Marc Lange offers a nice example backing Sober's claim. He says that claiming macro dominates micro ". . . would be like saying that it is likely for me to recover from my infection, given that I have been given the drug, since the Pr(recovery|drug) is high, even though I also know that I am one of those rare people on whom the drug is ineffective."[16] And this is a good example of a case where a relatively micro-chance should dominate a macro-chance. But my claim is simply that *sometimes* macro may dominate micro. Moreover, Lange's example (and many that Sober has in mind) are not ones where the micro-derived chances come from a micro-particle theory's stochastic laws, but rather ones where biological or medical sciences offer us two probabilities, one very coarse-grained and rough, the other more refined and sophisticated—but both in tune with the frequency patterns in the world.[17] These are not the sort of micro-derived chances whose primacy I am interested in challenging.

---

[16] Personal correspondence.

[17] I remind the reader that, as we saw in section 3.4 of chapter 3, HOC and PP do not lead to the counterintuitive advice to trust the more-coarse-grained chance over the more-fine-grained chance, if one knows both. Rather, knowing the latter gives an agent inadmissible information vis à vis the former chance, which means that the agent can rationally have credences not matching that more coarse-grained chance.

What about the superficially plausible idea that, while micro-derived chances may significantly diverge from actual frequencies *sometimes*, it is surely an occurrence both unlikely and infrequent? The plausibility is merely superficial. As far as "infrequent" goes, we could only give empirical evidence for this by comparing micro-derived chances with the macro-event frequencies, for real-world events. We would need the Laplacean demon's help to do that. And as for "unlikely," to make such a claim is a *petitio principii* of the first magnitude. "Unlikely" means "improbable"; that such deviations should be improbable on the chance side of things is presumably true, but irrelevant; what we need is "improbable" on the credence side. But the only way to establish *that* is to apply PP to micro-derived chances, the very sort of chance whose normative status is under dispute.

What makes the reliability of the micro-derived chances seem plausible is, perhaps, our tendency to think of these chances as *governing* events, or even as *producing* them. Thinking of our putative micro-chance laws the way a propensity theorist would, we fall easily into ascribing them an ontological importance that more macro-level chances cannot have. If the micro-chances are what *produce* the whole Humean Mosaic (bit by bit, over time), then surely whatever they imply has probability 0.66, *really does* have that probability? But a Humean perspective on objective chance is incompatible with thinking of chances in this way, i.e., as the makers of world history, guiding it in its unfolding over time. Chances are nothing but aspects of the patterns in world history. There are chance-making patterns at various ontological levels. Nothing makes the patterns at one level automatically dominate over those at another; at whatever level, the chances that can best play the PP role are those that count as the "real" objective probabilities.

Now, of course, if the micro-derived chances for generics like G are seriously divergent from the actual frequencies, then that is a strike against the theory's fit. We would expect that the Best System can't have too many such strikes against it—if any—because then it should lose the right to call itself best. Better fit could be purchased by complicating the former theory, telling the user (demon) not to do certain calculations, or not to apply PP to certain micro-derived probabilities. Earlier, when discussing the a priori justification of PP in connection with Q, I claimed that the improvement in fit gained by adding a special clause regarding Q would be nowhere near significant enough to offset the resulting loss of simplicity. But maybe this was too hasty a conclusion; if simplicity is given relatively little weight (as my preferred pragmatic approach to chance seems to entail), perhaps the truly "Best" system for our world would eliminate the frequency mismatch and salvage the applicability of micro-derived chances after all?

There are reasons to view this escape route as implausible. As we noted earlier, micro-derived chances for outcome-types of interest to humans are just a tiny fraction of the totality of outcomes involving huge numbers of particles that can be derived from a micro-level theory. As a purely mathematical fact, *some* outcome types will have a frequency significantly different from the derived chance: it is just logically inevitable, in a finite pattern at least. Therefore, no micro-level system can match the frequencies for all event types, all the time. Mismatching the frequencies of, say, breast cancers in city C in 2017 is just one of a huge number of mismatches that *any* micro-level theory is bound to have, unless equipped with a corresponding huge number of caveat clauses. However weakly we prize simplicity in our Best System competition, it is not going to give good marks to a system with an enormous number of caveat clauses.

In any event, saving a system from having micro-derived chances that fare badly in the actual course of events with caveat clauses does not alter the conceptual lesson of the whole discussion, namely: when micro-derived and macro-derived chances conflict, macro may well dominate micro, PP being correctly applicable to the macro chances but not the micro-derived chance. Caveat clauses do not solve this "problem," they illustrate it!

6.4. Summing Up

This chapter has been an exploration of some limitations that are conceptually built into the notion of HOC: limitations on the applicability of micro-level chances to guide credences for macro-level events. To those who (subconsciously or not) want the objective chances in our world to be *producers* and *explainers* of events as they unfold over time, these limitations will at first be unsettling. So it is worthwhile to recall how much they affect real human practices: *not at all.* We are not Laplace demons who can calculate probabilities for macro-scale events involving $>10^{23}$ particles. It is amusing to us to wonder what we should believe if we *could* do these things—because we are philosophers. But the limitations on the applicability of PP to Humean objective chances are philosophers' toys, not genuine limitations for science or any other human practice.

But we also saw in this chapter that even if we *could* perform the immense calculations needed to extract macro-level chances from micro-level chance rules, we would have no particular reason to trust them, and strong reason *not* to trust them, if they differed from the pronouncements of macro-level chance rules that we take to be plausibly part of the Best System. So micro-derived chances are irrelevant from a Humean perspective not merely for practical reasons, but also for theoretical ones.

7

Humean Chance in Physics

*Coauthored with Roman Frigg**

7.1. Introduction

After the acceptance of Newton's mechanics and the development of its increasingly sophisticated mathematical formulations, it was for over a century common to assume that determinism reigns in the physical world. Probability was no part of mainstream fundamental physical theory. Nevertheless, probability *was* introduced into physics during the height of the Newtonian paradigm, and its applications in physics—both at the level of theory, and in experimentation and instrumentation—have grown since by leaps and bounds to the point where it is hardly an exaggeration to say that probability is near-ubiquitous in modern physics.

In our view, the Humean approach to objective probability defended in this book is well suited to explicating the uses of objective probability that we have encountered in both theoretical physics (from quantum mechanics and statistical mechanics to modern stochastic-dynamics theories) and experimental physics (the various phases of experiment and equipment design, such as calibration and shielding techniques, data analysis, e.g., noise modeling, and so on). It is also well suited to understanding and justifying Monte Carlo techniques, which are used in many areas of physics practice. There is no way that we can survey all these applications in one chapter, however, and we will not make the attempt. Instead, we begin by observing that there are two basic ways in which objective probabilities seem to occur in physical theories: they can occur at a fundamental or "rock-bottom" level in the theory, or they can be *superimposed* on a deterministic fundamental

---

* *Department of Philosophy, Logic and Scientific Method and Centre for Philosophy of Natural and Social Sciences (CPNSS), London School of Economics*

level. In this chapter we will discuss paradigmatic examples of these two ways, starting with Boltzmannian statistical mechanics (SM), and then moving on to the standard interpretation of quantum mechanics (QM). These two cases, we believe, provide the template for understanding probabilities in other parts of physics that we cannot discuss here.

When probability was introduced in a quite crucial way into the physics of atomic and molecular systems, in Maxwell's and Boltzmann's works on statistical mechanics, it was possible to see its role as still essentially epistemic and non-fundamental, and this remains a viable view concerning classical statistical mechanics today (although not one we endorse). The rise of quantum mechanics in the 1920s famously changed this situation and turned things upside down: suddenly the *fundamental* physics of the micro-world was seen as genuinely indeterministic, governed by laws permitting only probabilistic predictions. Objective chances, it seemed, were both needed in *and provided by* fundamental physics; and it seemed that these chances, moreover, *could not be understood as merely epistemic*, in any of the ways familiar from nineteenth-century physics.

As we will see in this chapter, all the claims concerning probabilities in the preceding paragraph, for both SM and QM, are highly controversial and are still actively disputed in both the physics and philosophy of physics communities. But two facts are beyond dispute, and these facts make it highly desirable that we explore whether, and if so how, HOC can meet the challenge of delivering the objective probabilities of QM and SM. The first fact is this: both SM and QM (and later quantum field theories) have been enormously successful empirically and fruitful theoretically, and the probabilities they introduce are indispensable for these successes. The probabilities also generate hypotheses that are put to experimental test, and these hypotheses have been found to be in exact agreement with empirical results time and again. A natural explanation of this empirical success is that these probabilities capture "how things are," rather than just offering a codification of our own ignorance.

The second fact is that in both theories the probabilities are set in an "a priori" fashion, that is to say, derived from the theory's basic equations and models, not derived inductively from observation of frequencies or patterns of frequencies. This "*quasi*-a-priori" status is *prima facie* a challenge for a Humean account of all objective probabilities. For if objective chances are, so to speak, *written directly into the laws of nature*, this seems to challenge both the skeptical arguments of chapter 1, and the very need for a Humean/reductive account of probability (unless one is already committed to reductive Humeanism concerning the laws of nature in general, as is the case for some philosophers, but not for us). And even if SM probabilities are not taken to be directly written into the laws of nature, it is still desirable to show that the Best System recipe is likely to yield precisely the objective probabilities of SM (neither leaving them out of the picture entirely, nor amending them very significantly). In this chapter we will confront these challenges by exploring the ways in which HOC can be seen to capture both SM and QM probabilities.

7.2. Classical Statistical Mechanics

In this section we discuss the thorny problem of how to understand the probabilities introduced into physics by classical SM. Traditional approaches to probability, such as propensity views and frequentist accounts, run into serious difficulties, nor do special reductive definitions found in the theory (e.g., time averages) work at all well.[1] But as we will see, the pragmatic Humean approach seems to square nicely with the ontology and ideology of SM.

7.2.1. Background of Classical SM

The behavior of macroscopic systems such as a gas in a box is, to a very high degree of accuracy, correctly described by thermodynamics (TD). TD characterizes the states of systems in terms of variables like temperature, pressure, and volume, which pertain to the system as a whole, and it posits that processes have to be such that the entropy of

---

[1] Against a propensity interpretation of SM probabilities, see (Clark, 2001). Against the time averages approach, see (von Plato, 1981, 1982) and (van Lith, 2001).

an isolated system does not decrease. At the same time, the system can be regarded as a collection of molecules, each governed by the laws of mechanics, which in what follows we will (unless otherwise noted) assume to be the laws of classical mechanics (CM). Statistical mechanics aims to establish a connection between these two ways of looking at the system and to account for TD behavior in terms of the dynamical laws governing the microscopic constituents of macroscopic systems and probabilistic assumptions. We now briefly review classical SM and discuss in some detail how probabilities are introduced into that theory. SM comes in different versions and formulations. We here focus on what has become known as Boltzmannian SM and refer the reader to (Frigg, 2008), (Sklar, 1993), and (Uffink, 2006) for discussions of other approaches.

To introduce SM we first have to review briefly the main tenets of CM. CM describes the world as consisting of point-particles, which are located at a particular point in space and have a particular momentum (where a particle's momentum essentially is its velocity times its mass). A system's state is fully determined by a specification of each particle's position and momentum—that is, if you know the positions and the momenta of all particles in the system, you know everything that there is to know about the system's state from a mechanical point of view. Conjoining the space and momentum dimension of all particles of a system in one vector space yields the so-called phase space Γ of the system. For a particle moving around in the three-dimensional space of our everyday experience, the phase space basically consists of all points $X = (x, y, z, p_x, p_y, p_z)$, where x, y, and z are the three directions in space, and p_x, p_y, and p_z are the momenta in these directions. So the phase space of one particle has six (mathematical) dimensions. The phase space of a system consisting of two particles is the collection of all points $X = (x_1, y_1, z_1, x_2, y_2, z_2, p_{x_1}, p_{y_1}, p_{z_1}, p_{x_2}, p_{y_2}, p_{z_2})$, where x_1, y_1, and z_1 are the spatial locations of the first particle, x_2, y_2, and z_2 the one of the second particle, and $p_{x_1}, \ldots$, are the momenta in the respective directions. Hence, the phase space of such a system is 12-dimensional. The generalization of this to a system of n particles—which is what SM studies—is now straightforward: it is a $6n$-dimensional abstract mathematical space. If X is the state of an n particle gas, it is also referred to as the system's *micro-state*. So, the micro-state of a system consisting of n particles is specified by a point x in its $6n$-dimensional phase space Γ.

An important feature of Γ is that it is endowed with a so-called Lebesgue measure μ. Although Γ is an abstract mathematical space, the leading idea of a measure is exactly the same as that of a volume in the three-dimensional space of everyday experience: it is a device to attribute sizes to parts of space. We say that a certain collection of points of this space (for instance, the ones that lie inside a bottle) have a certain volume (for instance, one liter), and in the same way can we say that a certain set of points in Γ has a certain μ-measure. If A is a measurable set of points in Γ, we write $\mu(A)$ to denote the μ-measure of this set. At first it may seem counterintuitive to have measures ("volumes") in spaces of more than three dimensions (as the preceding X shows, the space of a one-particle system has 6 and that of a two-particle system 12 dimensions). However, the idea of a higher dimensional measure becomes rather natural when we recall that the moves we make when introducing higher dimensional measures are the same as when we generalize one-dimensional length, which *is* the Lebesgue measure in one dimension, to two dimensions, where the surface area is the Lebesgue measure, and then to three dimensions, where volume is the Lebesgue measure.

The state of a system will generally change over time in the way prescribed by classical mechanics. Since each particle moves along a continuous line in space, the system's micro-state in phase space similarly moves along some continuous path in phase space. The "line" that $\phi_t(X)$ traces through the phase space is called a trajectory. What trajectory a system follows depends on where the system starts. The state X where the system begins its motion at time t_0 (the moment when the process begins) is called the "initial condition." The system's trajectory is governed by the so-called *Hamiltonian equations of motion*. CM thus determines how a system located at any point in the phase space will evolve—move to a new point—as a function of time. The function that tells us what the system's state at some later point will be is called a "phase flow" and we denote it with the letter ϕ. We write $\phi_t(X)$ to denote the state into which X evolves under the dynamics of the system if time t (e.g., one hour) elapses, and similarly we write $\phi_t(A)$ to denote the image of a set A (of states) under the dynamics of the system.

More precisely, Hamilton's equations of motion define a *measure-preserving* phase flow ϕ_t on Γ. That is, $\phi_t : \Gamma \to \Gamma$ is a one-to-one mapping for every real number t and $\mu(\phi_t(R)) = \mu(R)$ for every measurable region $R \subseteq \Gamma$. In what follows, we assume that the relevant physical process begins at a particular instant t_0 and we adopt the convention that $\phi_t(x)$ denotes the state of the system at time $t_0 + t$ if it was in state x at t_0, and likewise $\phi_t(R)$; x is then commonly referred to as the "initial condition."

In (Boltzmannian) SM a key notion is that of a "macro-state." One assumes that every system has a certain number of macro-states $M_1, \ldots, M_k$ (where k is a natural number that depends on the specifics of the system), which are characterized by the values of macroscopic variables; in the case of a gas these would be pressure, temperature, and volume. It is a basic posit of Boltzmannian SM that the M_k supervene on the system's micro-states. Therefore each macro-state M_k is associated with a macro-region $\Gamma_k \subseteq \Gamma$, so that the system is in macro-state M_k at t iff its micro-state x lies within Γ_k at t.[2] For reasons that will become clear soon, we choose special labels for the first and the last macro-state: $M_1 = M_p$ and $M_k = M_{eq}$

In Figure 7.1 we see a schematic representation of the phase space of a system consisting of gas molecules in a box. Initially the system's molecules are all confined to the left side of the box. At time t_0 the partition is removed and the gas expands to fill the whole box; that is, its micro-state evolves from someplace inside M_p to end up somewhere in M_{eq}, the equilibrium macro-state. This evolution of the system takes it from a low-entropy initial state (in the instant after the partition is removed) to a high-entropy state. Showing that such an entropy-increasing evolution is overwhelmingly likely is a key aim of SM.

In TD, entropy is a notion defined in terms of macroscopic features of a system, so it makes sense that in SM entropy is defined for our

---

[2] Although thermodynamic properties such as temperature and pressure are continuous rather than discrete quantities, it is useful in SM to treat them in a "coarse-grained" fashion, so that a specific integer k indicates systems having temperature within a certain small range of values, pressure within a certain small range of values, etc. This gives the macro-regions of phase space corresponding to the macro-states M_k non-trivial volume, and is useful in other respects as well. See (Frigg, 2008) for details.

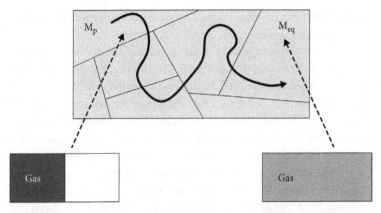

Figure 7.1. A micro-state starting in M_p, far from equilibrium, evolves through various intermediate macro-states and finally reaches M_{eq}, the equilibrium macro-state.

macro-states M_k. The Boltzmann entropy of a macro-state is defined as $S_B(M_k) = k_B \log(\mu(\Gamma_k))$, where k_B is the Boltzmann constant. Since the system is only exactly in one macro-state/macro-region at any moment, we can define the Boltzmann entropy of the *system* at time t simply as the entropy of the macro-state at time t: $S_B(t) = k_B \log(\mu(\Gamma_t))$, where Γ_t is the macro-region in which the system's micro-state is located at time t. It can be shown that, unlike the depiction in Figure 7.1, the equilibrium state M_{eq} is *by far* the largest of all states (under μ), a fact known as the dominance of the equilibrium state. In fact, for large n it is vastly larger than the area of all other regions.[3] Since the logarithm is a monotonic function, it follows that the Boltzmann entropy is maximal for the equilibrium macro-state.

The Second Law of TD says, very roughly, that the TD entropy of a closed system cannot decrease. In SM it is common to give a somewhat stronger reading to this law and take it to assert that a system, when left to its own, approaches equilibrium (and hence its entropy

---

[3] The notion of being large can be explained in different ways and different justifications can be given. For a discussion of this point see (Werndl & Frigg, 2015).

approaches its maximum value). Explaining this approach to equilibrium, then, is the aim of SM.

However, this aim is only of any interest if the system originally starts off in a non-equilibrium state. That this be the case is the subject matter of the so-called *Past Hypothesis*, the postulate that the system came into being in a low entropy state, which we call the *past state*. (The term "Past Hypothesis" is sometimes reserved for approaches in which the system under consideration is the entire universe, in which case it says that the universe came into being in a low-entropy macro-state, which is provided to us by modern Big Bang cosmology. We return later to the issue of the nature of systems studied in SM. In the meantime we use "Past Hypothesis" as referring to the initial state of a system, irrespective of the precise nature of the system.) Since the past state and the equilibrium state are of particular importance, we introduce special labels and denote the former by M_p (which is associated with Γ_p) and the latter by M_{eq} (which is associated with Γ_{eq}).

7.2.2. Explaining the Approach to Equilibrium

There are two different schools of thought that have divergent understandings of what exactly an explanation of the approach to equilibrium amounts to, and, accordingly, propose different solutions. One approach, the *TD-likeness approach*, will be explained in the following. The other, which we might call the *transition probabilities approach*, is associated with the work of Albert (2000). For the sake of brevity we concentrate on the first approach; readers are referred to (Frigg 2008) and (Frigg & Hoefer, 2015) for our take on the second approach.

The TD-likeness approach has in recent days been advocated by Lavis (2005) and can be traced back to Boltzmann himself. According to this approach, a justification of the second law amounts to showing that the system is highly likely to exhibit thermodynamic-like behavior (TD-like behavior). We have TD-like behavior *iff* the entropy of a system that is initially prepared in a low-entropy state increases until

it comes close to its maximum value, and then stays there and only exhibits frequent small and rare large fluctuations away from equilibrium (as depicted in Figure 7.2).

Before elucidating the notion of "highly likely" we have to introduce the concept of ergodic motion. Roughly speaking, the motion of a system is ergodic if for any set $A \subseteq \Gamma$, the proportion of time the trajectory spends in A is equal to the proportion of the measure A takes up in Γ in the long run (for instance, if A occupies a quarter of Γ, then the system spends a quarter of the time in A). If a trajectory is ergodic, then the system behaves TD-like because the dynamics will carry the system's state into Γ_{eq} and will keep it there most of the time. The system will move out of the equilibrium region every now and then and visit non-equilibrium states. Yet since these are small compared to Γ_{eq} it will only spend a small fraction of time there. Accordingly, the entropy is close to its maximum most of the time and fluctuates away from it only occasionally.

As a matter of fact, whether or not a system's motion is ergodic depends on the initial condition: some initial conditions lie on trajectories that are ergodic, while others don't. This realization is the clue to introducing probabilities. Consider an arbitrary subset $C \subseteq \Gamma_p$. We may postulate that the probability that the initial condition X lies within C at time t_0 is

$$p(C) = \frac{\mu(C)}{\mu(\Gamma_p)} \qquad \text{(Eq. 2)}$$

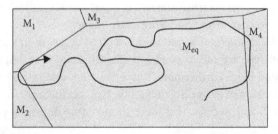

Figure 7.2. Time evolution of a system in equilibrium. It remains in M_{eq}, aside from brief sojourns into nearby non-equilibrium macro-states.

190

CHANCE IN THE WORLD

Let us refer to this principle as the Past Hypothesis Proportionality Postulate (PHPP). If we now denote by $E \subseteq \Gamma_p$ the subset of all initial conditions that lie on ergodic trajectories, then $p(E)$ is the probability that the system behaves TD-like:

$$p(\text{TD-like behavior}) = p(E) \qquad \text{(Eq. 3)}$$

If the values of $p(E)$ come out high, then one has justified the Second Law. Whether or not this is the case in actual systems is a non-trivial and controversial question. For the sake of argument let us assume that it is.[4]

7.2.3. Humean Chances in SM

We now argue that SM probabilities can be interpreted as HOCs. We begin to see how such chance and determinism are compatible if we pay attention to the curious hybrid character of Equation 3. The term on the right-hand side comes from the fundamental micro-theory, namely classical mechanics. The term on the left-hand side refers to probabilities for macroscopically observable events; in fact, we can sometimes see with our unaided eyes whether a gas in a container behaves thermodynamically. Given this, why do we need microphysics at all to attribute probabilities to these events? We assign probabilities to everyday events like getting heads when flipping a coin, seeing the sun shine tomorrow, cracking the jackpot in Sunday's lottery, seeing our favorite team win the next game, etc., without ever making reference to an underlying micro-theory.

Take the example of a coin toss, discussed at length in chapter 3. The event-type we call "a good flip of a fair coin" is widespread in HM around here. Furthermore, it is a fact, first, that in HM the relative frequency of each discernible side-type landing upward is very close to 0.5 and, second, that there are no easily discerned patterns

---

[4] In a recent paper, Roman Frigg and Charlotte Werndl (2011) argue that it is.

to the flip outcomes (it is not the case, for instance, that a long sequence of outcomes consist of alternating heads and tails). Does a Best
System for our HM contain a rule assigning probabilities to coin-flip
outcomes, and if so, what probabilities will the rule assign? The second
question is easy to answer: the rule that postulates $p(H) = p(T) = 0.5$
is the one that has best fit and simplicity. But does such a rule belong in the system at all? *Prima facie* the answer should be "yes," since
adding such a rule extends the system to a new and widespread type
of phenomenon in HM, at a tiny cost in reduced simplicity. As we
noted in chapter 3, there may be an even better chance rule that could
be part of the Best System, which would embrace coins and dice and
tetrahedra and dodecahedra and other such symmetric, flippable/roll-
able solids. The rule would say that where such-and-such symmetry is
to be found in a solid object of middling size with n possible faces that
can land upward (or downward, thinking of tetrahedra), and when
such objects are thrown/rolled, the chance of each distinct face being
the one that lands up (or down) is exactly $1/n$. Given what we know
about dice and tetrahedra and so forth, it is quite plausible that this
rule belongs in the Best System; and it entails the coin-flip chances. So
it enhances both simplicity and strength without much loss in fit, and
hence on balance it is better to add this rule to the system rather than
a set of rules, one for each type of regular n-sided solid. Hence, the
chance of heads on a fair flip of a coin would seem certainly to exist,
and be 0.5, in a Best System for our world.

The same kind of reasoning applies to gases. Behaving TD-like is a
macro-property in much the same way as showing heads, and so we
can introduce chances for that type of event in the same way. Preparing
a gas in M_p and then letting it evolve unperturbed corresponds to flipping a coin; the gas behaving TD-like or not correponds to getting
heads or tails. HM contains many gases that were first prepared in M_p
and then left to themselves and so we have solid frequency data to rely
on (just as in the case of coins). The overwhelming majority of those
behave TD-like and so we formulate the rule $p(\text{TD-like}) = 0.9999$ and
$p(\text{non-TD-like}) = 0.0001$ (omitting many 9s and 0s here for brevity).
Let us refer to these as the *macro-probability rules* for TD-like and
non-TD-like behavior, respectively. This rule is simple and has good
fit. It also possesses strength because it turns out that it holds true not

only for gases; in fact, liquids and solids also behave TD-like most of the time and so the rule is applicable universally (or almost universally at any rate). For this reason our rule is part of the Best System, and the probabilities it introduces are chances in the sense of HOC.

But notice that something curious has happened. We started explaining the hybrid character of Equation 3, and ended up making claims about the probabilities that appear in Equation 3 without reference to that equation at all! Equation 3 *seems* to have become an idle wheel. Has it really? The answer is "yes" and "no."

"Yes" because, and this is an important point, probabilities for macro-events like coin flips and the behavior of gases can (a) be defined without reference to the underlying micro-physics and (b) be genuine parts of the Best System. Especially the second part of this claim may strike some as unfounded, and we will come back to it later. Let us for now accept that macro-chances are genuine chances. The "no" part of the preceding answer holds that Equation 3 is not an idle wheel *even if* one accepts that macro-chances are real and genuine chances.

How can that be? The crucial point to realize is that, first appearances notwithstanding, Equation 3 does not *give* us the chance for TD-like behavior. We don't need to be given anything—we have the chance (via the *macro-probability rule*). Rather, Equation 3 is both a consistency check and an explanation. We don't want to place too much emphasis on the latter and mainly focus on the former, but there is the pervasive intuition that if a macro-result can be derived from a more fundamental theory, there is explanation. Those who share this intuition—among them us—can see Equation 3 as providing an explanation of the *macro-probability rule* for TD-like behavior (because set *E*, which is the key ingredient of that equation, is in fact given to us by the dynamical hypothesis that the system be ergodic).[5]

---

[5] The character and the strength of the explanation provided depends, of course, on how one views several things, including: the fundamentality of SM; the status of fundamental mechanical laws (whether of CM or QM); and the success of ergodicity-based explanations of thermodynamic behavior. For an account of the latter item, see (Frigg & Werndl 2011).

Let us now turn to consistency. The different parts of a Best System have to be consistent with each other. For this reason, whenever a macro-level chance rule and a micro-level chance rule are extensionally equivalent, then the chances they ascribe must agree, or be very nearly in agreement. This, of course, does not rule out the possibility of minor adjustment. For example, assume we adopted the 50/50 rule for heads and tails when flipping a coin. Now suppose we know for sure that we get the reductive relations right and we have the correct micro-theory, and based on these we find 49/51. This is no real conflict because there is some flexibility about the macro-chances, and if there are very good overall reasons for making adjustments, then the Humean can make these. But there is a breaking point: if the micro-theory predicts 80/20, we have to go back to the drawing board.

Recall Equation 3: p (TD-like behavior) = $p(E)$. This equation provides a consistency check in two ways. The first and more straightforward one is that a calculation has to show that $p(E) = \mu(E) / \mu(\Gamma_p) = 0.9999$, or at least that the value of $p(E)$ is very close to 0.9999. If the system is ergodic we find that $p(E) = 1$ because initial conditions that lie on trajectories for which time and space averages do not coincide form a set of measure zero. So this requirement is met. The second way has to do with the supervenience of the measure in Equation 3. It is one of the posits of HOC that chance functions supervene on the Humean Mosaic. But before we can see what consistency constraint emerges from that posit, more needs to be said about what it means for a rule like Equation 3 to supervene on the HM.

To understand what it means for Equation 3 to supervene on the HM, we first have to ask the same question about Equation 2—i.e., $p(C) = \mu(C) / \mu(\Gamma_p)$. To begin with, notice that this equation in effect expresses the conditional probability of finding a system's microstate in set C given that it lies in Γ_p (because by assumption $C \subseteq \Gamma_p$). Now look at the same conditional probability from the point of view of HOC. There is a well-circumscribed class of objects to which a chance rule like PHPP applies (gases, etc.). Each of these, we are assuming here, is a classical system with a precise initial condition X at t_0, which, by assumption, lies within Γ_p. Now go through the entire HM and put every single initial condition x into Γ_p. The result of this is a swarm of points in Γ_p. Then recall that HOC can be viewed as a sophistication of finite frequentism, and chances should closely track relative frequencies wherever such frequencies are available in large quantity. Hence the chance of an initial condition being in set C given that it lies in Γ_p should be close to the fraction of points in set Γ_p that lie in C (in the same way in which the chance of heads is the fraction of heads in the set of all coin toss outcomes).

But listing all points individually and checking whether they lie in C is extremely cumbersome and won't make for a simple system. So we have to reduce the complexity of the sytem by giving a simple summary of the distribution of points. To this aim we approximate the swarm of points with a continuous distribution (which can be done using one of the well-known statistical techniques) and normalize it. The result of this is a probability density fuction ρ on Γ_p, which can be regarded as an expression of the "initial condition density" in different subsets C of Γ_p. This distribution supervenes

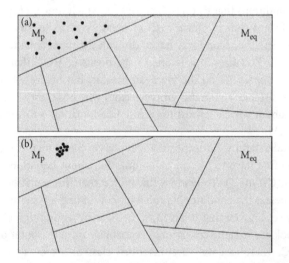

Figure 7.3a,b. Actual initial conditions of systems that start out in M_p. If evenly distributed (*a*), the Lebesgue measure will provide a good probability measure, as in Equation 2. If the actual initial conditions are too clustered (*b*) this will not generally be the case.

on the HM by construction. The consistency constraint now is that $\rho(C)$ be equal to (or in very close agreemet with) $\mu(C) / \mu(\Gamma_p)$ for all subsets C of Γ_p. This is a non-trivial constraint. For it to be true, it has to be the case that the initial conditions are more or less evenly distributed over Γ_p because $\mu(C) / \mu(\Gamma_p)$ is a flat distribution over Γ_p. This is illustrated in Figure 7.3 a. If it turned out that all initial conditions were crammed into one corner of Γ_p, as in Figure 7.3 b, then Equation 2 would have poor fit, which would preclude its being part of the Best System, despite its great simplicity. So the requirement $\rho(C) = \mu(C) / \mu(\Gamma_p)$ presents a real touchstone for the system. Equation 3 is then dealt with easily. Since Equation 2 has to hold for all C, a fortiori it has to hold for E. Hence Equation 3 gives the correct chance for TD-like behavior.

This leaves us with the question of why the Best System for our world might have macro-chances for TD-like behavior, if the SM chances emerge as part of the system in the way sketched in the preceding. Would they not be superfluous?

We think not. Even if the world is classical at bottom and classical mechanics is the fundamental theory of the universe, it does not follow that everything that can be said about HM has to be said in the language of the fundamental theory. As argued in chapters 3–5, probability rules can be formulated in a macro-language pertaining to a certain level of discourse, and probability rules thus introduced have equal right to be considered for inclusion in a Best System package of rules, alongside micro-level rules. On our view, then, the $1/n$ rule for gambling devices and the *macro-probability rule* for TD-like behavior are genuine chance rules because they improve the score, in terms of the three basic metatheoretical virtues, of a system that includes them, compared to a system that does not and (say) only contains micro-level chance rules.

The key point is that our pragmatic Best System account characterizes *simplicity* in a way that connects with the way real sciences work. A system that has rules at various levels and for various domains, not just micro-physical-level rules, is one that has more relevance to real science. It is a matter of fact that different scientific disciplines talk in their own language, and we have no reason to believe that this will change. Biology speaks its own language, and there is no indication that progress in biology would mean that biologists start formulating their theories in the terms of fundamental physics. A view of chance (or laws of nature, for those who adopt a HBS view of laws too) that ignores this aspect of science is one that is of little use to science.[6]

One might worry that adding probability rules in a macro-language to the system in fact renders the system less simple without adding any strength, at least if the system already has micro-level rules that extensionally "cover" all the phenomena covered by the macro-level rule (as one might say when comparing Equation 3 with the macro-probability rule introduced earlier). This is mistaken. In fact, adding these macro-level rules can make the system simpler! The reason for this is that aspect of simplicity mentioned in earlier chapters, simplicity in derivation: it is hugely costly to start from first principles every time you want to make a prediction about the behavior of a macro-object. So the system becomes simpler in that sense if we write in rules about macro-objects.

Some philosophers would object that the chance for macro-level event types cannot be independent of the micro-physics of the world. Surely, the argument goes, there must be some dependence there! If the physics of our world were vastly different from what it is, then the chance for coin flips to land heads would be different too. So one cannot just introduce chances at a macro-level and completely disregard the fundamental physics.

There is a grain of truth in this, but we must not be misled. First, the physics of our world might be vastly different and yet (for whatever reason) the pattern of heads- and tails-outcomes in HM might be exactly the same; in which case, the chances could be the same. Imagine a universe in which matter is a continuum and obeys something like the laws of Cartesian physics; imagine that gases exist in such a world and spread in the way we are used to. Despite the basic physics being very different, the chance for TD behavior could be

---

[6] Notice that the replacement of one language by another is not even a requirement for reduction, which only requires that the terms of the two theories be connected by bridge laws. For a discussion of this aspect of reduction see (Dizadji-Bahmani, Frigg & Hartmann, 2010).

the same because the overall pattern of gases spreading in HM might be the same. Second, insisting, as we do, that macro-level chances are genuine chances does not imply that such chances be logically independent of chance rules or laws at the micro-level; i.e., does not imply that such chances cannot potentially conflict with rules and laws at the micro-level. There can be a clash between rules, and as we have seen earlier, we do want macro- and micro-rules to be consistent (we interpreted Equation 3 as a consistency constraint!). But this does not imply that macro-chances are redundant, or that they aren't chances at all (as we saw in chapters 3 and 6, specifically sections 3.4 and 6.2.2).

Residual unease about macro-probability rules and macro-chances might come from the common intuition that there is something epistemic about our probability for a gas to behave TD-like—after all, the gas has one and only one initial condition, and given this initial condition it is determined whether or not it behaves TD-like. This worry can be defused by recalling the discussion of chance and determinism from chapter 3. Information about the precise initial condition (IC) of a particular gas is certainly *inadmissible*, for application of PP, for an agent whose background knowledge includes the micro-level laws: together, such information logically implies how the gas will behave and hence provides (in principle!) knowledge about the gas's behavior that does not come by way of information about the HOC chances. The crucial point is that in typical situations in which we observe gases, we just don't have the inadmissible IC information, or the ability to calculate what 10^{23} particles will do from a given IC, and that is why we use chances and PP to set our degrees of belief. So we use chances when we lack better knowledge. But this does not turn HOCs into Bayesian ignorance probabilities. What it shows is that demons who can observe precise initial conditions and do calculations fast enough have no use for HOCs; they have better information with which to guide their credences about future events than the information HOCs provide. We humans, alas, never have had nor will have either such information about initial conditions, or such demonic calculational abilities. For us, it is a good thing that objective chances exist, and that we can come to know (and use) them.

7.2.3. Summing Up SM

It is plausible that a Best System for our world will capture the probabilities of Boltzmannian SM. In fact, we suggest that it may do so twice over, so to speak: first, in terms of strictly macro-level probabilities for macro-state transitions or for TD-like behavior; and second, by including a chance rule corresponding to the PHPP. Whether the latter truly deserves to be part of the Best System, and is compatible with the former in the sense demanded by HOC, is something that cannot be proven. But we feel that the successes in application of SM so far provide positive evidence for this.

7.3. Quantum Chances

7.3.1. "Standard" QM

As we noted in the introduction to this chapter, it was with the rise of QM that physicists first started to take seriously that the fundamental physics of our world might be irreducibly probabilistic, that is, able to provide at most probabilistic predictions concerning future events, even given a *complete* description of the state of affairs at a certain moment of time. While this turn away from determinism was unwelcome to many physicists, Einstein most prominently, others embraced this and the other radical oddities of the quantum theory that emerged in the 1920s. The radicals (Bohr, Heisenberg, Pauli, Jordan, and others) eventually won out over the skeptics,[7] leading to the establishment of a standard formulation of QM and an accompanying interpretation of the formalism that remain, nevertheless, controversial even today. The cause of continuing controversy is not the fact that QM introduces fundamental probabilities per se, but rather the way in which it does so: in the context of *measurements*.

---

[7] For discussion of the history of how the "orthodox" interpretation of quantum theory came to predominance in physics, see (Cushing, 1994).

Although we cannot give a complete introduction to the basic features of QM here, we will briefly recount some of the basic elements of standard QM and how they lead to the so-called measurement problem, which is crucial for an understanding of how probability does and doesn't function in standard QM.[8]

QM describes physical systems by specifying a mathematical quantum state, represented by a vector in some vector space. This vector space, often called a Hilbert space, permits the representation of all the possible physical states that the system might have, given the kind of system it is (e.g., electron, or proton, or a pair of electrons emitted from a single source, or an *He* atom, etc.). A curious and important feature of QM is that there is more than one way to represent a system using vectors in a vector space, and the different ways correspond to different *properties* that the physical system can be seen as having: properties such as *total energy*, or *momentum* (in some frame of reference), or *position* (again, in some coordinate frame).

The representation of microscopic systems in terms of *position* and *momentum* is important for many applications of QM. However, properties that can (in principle) take on a continuum of distinct values, such as position (and momentum) involve—even for quantum systems that consist of only a single particle—an *infinite*-dimensional Hilbert space, and bring with them certain mathematical complexities that are not relevant for understanding the core notions of QM and the measurement problem to which they give rise. So we will introduce QM with the example of the spin of a simple quantum particle, such as an electron or proton, which has only a two-dimensional Hilbert space.

Spin is genuine quantum property that has no real classical equivalent. The closest "classical equivalent" one can find is the angular momentum of a sphere rotating around an axis that goes through its center (like the rotation of the earth around its axis), where the magnitude of the angular momentum depends on the radius of the sphere, its mass, and the speed of rotation. But thinking of the quantum spin

---

[8] Excellent introductory texts to conceptual issues in QM are (Albert, 1992) and (Maudlin, 2019). For an in-depth yet non-expert introduction to the formalism of QM see (Hughes, 1989).

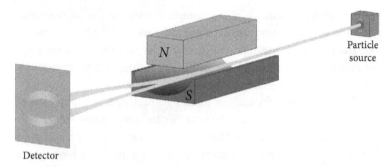

Figure 7.4. Stern-Gerlach device for measuring electron spin.

of an electron in classical terms is misleading because the electron is a point particle (and as such has no radius) and it makes little sense to say that a point rotates.[9] So it is best not to visualize spin and instead think of it in terms of its experimental manifestations. The most important of these manifestations is what we observe when an electron is placed in an external magnetic field. QM associates a magnetic moment with spin and hence when an electron is placed in a magnetic field it experiences a force. The effect of this force is observed in a famous experiment called the Stern-Gerlach experiment. Figure 7.4 shows a schematic representation of the experiment.

The core of the experiment is a magnet that produces an inhomogeneous magnetic field. A source (shown on the right-hand side) produces a beam of electrons that moves through the magnet. While in the magnet, magnetic moment of the spins interact with the magnetic field, either pulling the electrons up (if the north pole of their magnetic moment points downward) or pushing them down (if the north pole of their magnetic moment points upward). As a result, the electrons deviate from the straight path they would have taken in the absence of the magnetic field. A screen at the end of the magnet (shown on the left-hand side) registers where the electrons

---

[9] From the perspective of later quantum field theories, the electron is not a point particle, or indeed a particle at all, but rather a "quantum of excitation of the quantum field." In so far as we understand what this means, it seems clear that it remains true that it makes no sense to think of such a thing as rotating.

are. Let's assume we have a Cartesian coordinate system that is such that the beam moves along the x-axis and the vertical direction is the z-axis. Classically one would expect that the magnetic moments in the z-direction could have any magnitude, and so one would expect to find a vertical line on the screen. However, the experiment shows two points rather than a line. So all electrons are diverted either upward or downward by the same amount, and so in effect the beam gets split into two. There is never any other diversion, and in particular no zero diversion.[10] Moreover, the beams going up and down have the same intensity, and so we infer that the same number of electrons are diverted upward and downward. From these observations, one infers that the angular momentum of an electron in z-direction can have only two values (which have same magnitude but opposite signs), and that these values occur with the same frequency in the beam emitted by the source.

QM explains this experimental behavior of electrons by postulating that an electron can have two z-spin states, $|u\rangle$ and $|d\rangle$. The brackets indicate that the states are elements of a vector space, the system's Hilbert space, and the u and d are mnemonic notations motivated by the motion of the electron in the magnetic field: u stands for "up" and d for "down." If the electron is in state $|u\rangle$, it moves upward when moving through the magnetic field; if it is in state $|d\rangle$, it moves downward. This might suggest that the Stern-Gerlach experiment could be interpreted as a device that *reveals* the quantum state of an electron: the source produces electrons that are either in state $|u\rangle$ or $|d\rangle$, and the apparatus sorts the original mix of both into two beams, one in which all electrons are in state $|u\rangle$ and one in which all are in $|d\rangle$.

Unfortunately, this is not what happens and things are a bit more complicated—and these complications matter because the moves that QM makes to circumvent the complications is what brings the probabilities into the theory. Quantum states, as mentioned earlier, are elements of a vector space. This matters here because it implies that they can be multiplied with constants and added together, and

---

[10] This is so for particles with half-integral spin ($\pm 1/2$, $\pm 3/2$, $\pm 5/2$, etc.). Electrons and protons are such particles (fermions). Bosons, however, have integral spin and can have spin values of zero, which can lead to no diversion.

the result of such an operation is still a member of the vector space. Specifically, if $|u\rangle$ and $|d\rangle$ are in the vector space, then $|v\rangle = a|u\rangle + b|d\rangle$ is also in the vector space for any complex numbers a and b that satisfy the condition $|a|^2 + |b|^2 = 1$. For this reason, the electron cannot only be either in state $|u\rangle$ or $|d\rangle$; it can just as well be in any state of the form $|v\rangle = a|u\rangle + b|d\rangle$. In QM jargon, such a state is referred to as a "superposition." An electron being in a superposition is more than just a "theoretical," or even "mathematical," possibility—such states do occur in nature. In fact, it turns out that the electrons that are produced by the source are in a state like $|v\rangle$ (and there are experimentally established ways to *prepare* a stream of electrons so that they can be known to be in a state like $|v\rangle$, if one is unsure about their initial state. (For a detailed account of one way to do this, see (Albert, 1992), Ch. 2).

But now we are in trouble. What outcome are we to expect in the Stern-Gerlach experiment if the electron enters the magnet in a state like $|v\rangle$? Our current interpretation does not cover such cases, and there does not seem to be room for states like $|v\rangle$ in the account as developed so far. If the electron is in state $|u\rangle$ it moves up and if it is in $|d\rangle$ it moves down, and these are the only two things that happen in the experiment. So there is no experimental outcome we could assign to $|v\rangle$. This problem is resolved by the infamous *Collapse Postulate*, which in the current context says that when z-spin measurement is performed, then the state of the electron collapses either to state $|u\rangle$ or $|d\rangle$, no matter what the electron's state prior to measurement is. That is, the postulate says that if an electron in state $|v\rangle$ enters into the measurement device, the electron will be either in state $|u\rangle$ or $|d\rangle$ when the measurement is concluded. So the z-spin measurement device changes the electron's state to a state that is interpretable in terms of z-spin properties!

The remaining question is which of the two possibilities is actualized. What standard QM says is that the outcome is not determined in advance, but is rather a matter of chance. This is codified in the *Born Rule*, which assigns probabilities to the different possible outcomes of a measurement. In our context the rule says that if an electron in state $|v\rangle = a|u\rangle + b|d\rangle$ enters the measurement device, then the probability of finding the electron in either of the two z-spin states is equal to the square of the coefficient of this state in $|v\rangle$. Specifically, the probability of finding an electron in state $|u\rangle$ is $|a|^2$ and the probability of finding it in state $|d\rangle$ is state $|b|^2$. So if the electrons enter the system in state $|s\rangle = \sqrt{1/2}|u\rangle + \sqrt{1/2}|d\rangle$, then each electron has a probability of 1/2 to come out in state $|u\rangle$ and a probability of $1/2$ $|d\rangle$. A beam contains an extremely large number of electrons, and for this reason about half of the electrons end up in state $|u\rangle$ after the measurement, and the other half in state $|d\rangle$. This explains the splitting of the beam into two rays of equal intensity.

It's worth emphasizing that the preceding is merely an example, and the two rules we have applied to the case—the Collapse Postulate and the Born Rule—are completely general. Assume we perform a measurement of property P, and let us call "P-states" the states that can be interpreted as giving definite P-outcome (in the previous example P is z-spin, and $|u\rangle$ and $|d\rangle$ are z-spin-states). Whenever we perform a measurement of property P and the system's state prior to measurement is in a superposition of P-states, then the state collapses onto one of the P-states with a probability equal to the square of the coefficient of this state (in our example the probabilities are $|a|^2$ and $|b|^2$, respectively).

These two rules are immensely effective in generating successful predictions. To date no experiment is known that contradicts QM as formulated in the preceding, and it is the version of QM that working physicists operate with. For this reason we call it *standard* QM. But qualifying a theory as "standard" is meaningful only if there are "non-standard" versions of it. And there are. To understand what these versions are, we need to articulate the (in)famous *measurement problem* of QM, which arises in the preceding theory. While all versions of QM retain (some version of) the Born Rule, the collapse postulate is highly controversial and the main aim of non-standard versions of QM is to exorcise the postulate.

We cannot review alternative versions of QM here, but we would like to explain at least briefly why the Collapse Postulate is widely considered to be problematic. Like CM, QM has dynamical rules (or laws) describing how the state of any system should evolve over time.

In CM states evolve according to Newton's (or Hamilton's) equation of motion, and in QM they evolve according to Schrödinger's equation of motion. QM is a general theory and so its rules for the time evolution of states in principle cover how *any* material system should evolve over time; the system could be a single particle subjected to some potential, or a set of particles, or even a huge collection of particles. A fortiori they also cover what happens when a pair of distinct systems, initially separate from each other, come into interaction with each other. A particular case of such an interaction is when a small quantum system (such as an electron) interacts with a large multi-particle system (such as a Stern-Gerlach magnet); that is, the kind of interaction we tend to think of as a *measurement*. Now, an essential mathematical feature of the dynamical laws of QM is their "linearity." We need not worry about how linearity is defined here, because its consequence is easy to state non-mathematically. When a small quantum system in a *P*-property superposition state interacts with a *P*-measurement device, what the dynamical laws entail is that the superposition should "infect" the device: the combined system should evolve into one in which the small quantum system is still not in any definite *P*-property state, and the device is in a superposition of distinct measurement-result states! In other words, what the dynamical rules alone entail is that measurements do not have single results, but rather *all* the possible results. The quantum description of the combined system becomes a superposition, and there is no such thing as "the result that actually occurs." This does not, however, appear to be what really happens when we do measurements in our labs with things like a Stern-Gerlach apparatus. Reality appears to keep our Stern-Gerlach apparati in one definite state, corresponding to one or the other of the possible *z*-spin measurement results.

One version of the measurement problem, then, is that the dynamical laws of QM predict no definite outcomes, and instead predict that macroscopic objects should evolve into superposition states, but this does not appear to correspond to our experience. But the reader has already seen the solution to this problem that standard QM offers: the Collapse Postulate! To use the theory in the context of understanding measurement interactions, one sets aside the standard laws of dynamical evolution at the point where one considers a measurement to have occurred, scratches out the superposition, and rewrites the quantum state of the system as the definite-P-property state corresponding to the observed measurement outcome. Similarly, the quantum description of the measuring device (if one were interested in thinking about it, which usually one is not) should be not a superposition, but a nice definite-P-property-registering state. The collapse rule saves QM from giving us an apparently wrong description of what happens in our laboratories, and allows the theory to make exceptionally well-verified predictions regarding measurements of all sorts.

The price of this solution of the measurement problem, however, is high—for many, too high to be borne. This solution involves introducing an *epistemic* notion, *measurement*, into the dynamical laws of a fundamental theory in an essential way. From a physical standpoint, a measurement device is just a many-particle system, no different from any other system with a similar number of atoms. A puddle or a vapor cloud can deviate the path of a flying electron, just as a Stern-Gerlach apparatus does; what makes the latter interaction dynamically special? For that matter, what *is* a measurement? Which interactions count as measurements and which do not—and *why*? Although it is clear enough how to apply QM in all practical situations, the need to introduce the notion of measurement into the fundaments of the theory leaves it ill-defined in a way that most other physical theories are not, leading many philosophers and physicists to look for some revised interpretation of the theory, or even a replacement for it, that removes the need to talk about measurements as a special kind of interaction. To date, no alternative to the standard theory has gained anything close to acceptance by a majority of physicists (or philosophers). For this reason, and to keep the discussion manageable, we discuss probabilities as they occur in standard QM.

7.3.2. Standard QM and HOC

Thus, in the 1920s a remarkable situation came into being, one that persists to this day: our best fundamental theory of ordinary matter does not give us a coherent, realistically understandable *description* of what matter is and how it behaves, but rather only an instrumental *prescription* for making predictions—probabilistic predictions, in the main—about the results of observations and experiments. As Tim Maudlin so eloquently argues in his (2019) introduction to QM, one should really not think of QM as a *theory*, but rather as a *recipe* for making certain sorts of predictions.

And as everybody knows, this instrument, this recipe, is fantastically successful: the patterns of measurement outcomes in actual events that constitute quantum experiments are just as you would expect given the objective probabilities generated by the Born Rule. Another way to put this fact: the Humean Mosaic of our world appears to have certain widespread, reliable stochastic-regularity patterns in it, patterns that can be captured and systematized with great strength, amazing simplicity, and fantastically good fit, by the axioms of standard QM including the Born Rule. In other words: prima facie, standard QM is a strong candidate for inclusion in the Humean Best System for our world, *even* (nay—especially!) *if taken as nothing more than a recipe for predicting events that are all, in the end, human observation-events.*[11]

This is so, however, only for a pragmatic HOC such as we advocate in this book, and apparently not for Lewis' Best System theory, because in the latter the laws are supposed to only invoke the *perfectly natural properties* (natural kinds) from which events in the fundamental HM are composed. Whatever account of "perfectly natural properties" one may offer, the property of *being a measurement* is presumably not going to count as perfectly natural. By contrast, a pragmatic approach such as ours, which is happy to look for patterns in the HM at various "levels," expressible with a wide range of predicates and terms that have proven useful for human beings—trains and tables as

[11] By judiciously ignoring the quantum measurement problem, the apparent limitation of this recipe to predicting observations made by humans and other rational beings can be pushed out of sight, and one can then talk indiscriminately about events such as radioactive atoms decaying, electrons detaching from an atom by quantum tunneling, and so forth, independent (apparently) of whether measured/observed or not. Physicists do this all the time, and it is "cheating," for reasons well understood in the philosophy of QM literature. But even if we avoid cheating and stick to merely systematizing actual human measurements and observations that are covered by the QM recipe, there is still an awful lot of stuff in the HM that gets simply and strongly systematized by a Best System that includes the QM rules.

well as quarks and electrons—need not blush at incorporating a rule such as the Born Rule, even if it refers to a kind of event that can only be understood as the activity of an epistemic agent.

So we have eminently good reason to think that a Best System of our world, under the pragmatic account of HOC defended in this book, would include the Born Rule as an element, with the rest of the theory of standard QM written into the system as auxiliary content necessary to specify the chance setups in which the rule applies and to determine the mathematical values of the probabilities. There is, moreover, no reason why this should be restricted to the non-relativistic QM that emerged in the late 1920s. Quantum field theories, which now are systematized in the Standard Model of "particle" physics,[12] share all the relevant features of QM just discussed, and would deserve to be parts of the Best System for exactly the same reasons.

Can alternative views of objective chance also make good sense of QM probabilities? As we just noted, on the face of it, Lewis' account is difficult to square with the Born Rule's invocation of *measurement*.[13] For similar reasons, chance primitivists ought to feel uncomfortable with standard QM as well: why is it *measurement situations* that have associated objective chances for outcomes, rather than, say, *interactions of big/many-particle systems with little/few-particle systems*? The latter is awfully vague, but at least it is not overtly epistemic. And many physicists and philosophers of physics have speculated that something like this must really be the better description of where nature goes indeterministic.

The idea would be that the big-system/little-system interaction somehow provokes an objective "collapse" of the quantum state of the little system, taking it from superposition to eigenstate (or at

---

[12] We put "particle" in scare quotes here because the majority view among physicists is that *fields* are fundamental rather than particles. We take no sides in this still-contentious debate.

[13] Barry Loewer's version of the Best System theory of laws, however, is explicitly more pragmatic in its approach, and might be able to lay claim to standard QM as a likely part of the Best System for our world. I suspect, however, that Loewer would prefer to see some replacement version of QM (e.g., some relativistic successor to Bohm theory or GRW) built into the Best System. Although more pragmatic than Lewis, Loewer is still aiming to offer an account of the true fundamental physical laws.

least "sharply localized" states, for continuous variables such as position). There are problems with this idea. First, it is no good if left in the vague form indicated; to be a worthy proposal, it would have to be cashed out in some concrete and precise fashion. This would make the resulting theory, in principle, at least *slightly* different from standard QM in its predictions, since standard QM involves no spontaneous collapses and instead says that the superposition of the little system can turn into a superposition of the combined big-system/ little-system system (this being the lesson of Schrödinger's cat). It would be a new theory, in other words, and not standard QM. Second, experiments that have gone looking for spontaneous collapses have so far not found any. Nature so far appears stubbornly fond of the instrumentalist QM recipe (though it is clear that, depending on how the spontaneous collapses work, it may be near-impossible to design an experiment that would reveal them). This is the case for the GRW theory, which is precisely a well-worked-out spontaneous collapse theory of the type we are considering here; for a discussion of this theory and how HOC may work for it, see (Frigg and Hoefer 2007).

So standard QM is not a very apt home for primitive chances or chancy fundamental laws, but some nearby, reformed theory (such as GRW) might look like a natural candidate for a primitive-chance reading. That is, it might look that way at first blush; but the arguments of chapter 1 aimed to convince the reader that this interpretation is not nearly as clear and unproblematic as it may initially seem. Readers who feel attracted to such a reading of QM probabilities are invited to re-read section 1.3.

7.3.3. Humean Chance and the Wave Function

We end with a final note about how HOC squares with standard QM more easily than a strict Lewisian Best System approach does. In our introduction to the main features of QM in the preceding, we chose to look at spin properties and states; these are easy to represent and to use to illustrate the notions of superposition and measurement. We did not look at how QM represents either the *position* of particles and or their *momenta*. Position and momentum, unlike spin, are variables that range over a continuum of possible values, and quantum states of position or momentum are represented by continuous functions—so-called *wave functions*—defined over the continuous spaces that codify all the possible particle positions (or momenta—but from now on, we will set aside momentum and just think about position) of the quantum system.

To be more precise: the wave function ψ of a system is a complex-valued field defined in the so-called *configuration space* of the system.[14] The notion of "configuration space" is borrowed from CM, and is very similar to the notion of phase space we saw in section 7.2. In CM, if we have a system of N particles, it is sometimes convenient to represent the positions of all the particles through a unique point $Q \equiv (\boldsymbol{Q}_1, \boldsymbol{Q}_2, ..., \boldsymbol{Q}_N) \in \mathbb{R}^{3N}$, where $\boldsymbol{Q}_i \in \mathbb{R}^3$ are the position coordinates in physical three-dimensional space of the ith particle. Configuration space is the set of all points that—like Q—represent a possible configuration of all the particles of the system in physical three-dimensional space, and it trivially follows that configuration space is $3N$ dimensional.[15]

In QM, the fact that the wave function ψ of a system is defined in configuration space implies the following. If we have a one-particle system, its wave function assigns a number to each point of the ordinary, three-dimensional space. In this respect, the wave function of a one-particle system can be thought of as analogous to a classical field: at each point in space, the wave function has a certain numerical value, or *amplitude*. If we have a two-particle system, however, then six spatial coordinates—not just three—must be used in order to specify the wave function. And in general, for an N-particle system, $3N$ spatial coordinates are required to specify the wave function.

---

[14] In other words, the wave function assigns a complex number to each point of that space. The fact that the values of the wave function at points (also called the *amplitude* at that point) is a complex number is a detail that we will ignore from now on.

[15] A difference between configuration space and phase space is that the phase space for a system of N particles has $6N$ dimensions, because a point in phase space represents not just the instantaneous spatial configuration of all the particles, but also their velocities or momenta.

A *point* in configuration space represents each of the N particles having some *specific* location in ordinary 3-space. (And a continuous line in configuration space could by extension represent how N particles all change their locations in space, over time.) But the wave function does not select a point of configuration space; it assigns some field value, or amplitude, to every point in configuration space. How does such a thing *represent* the state of a system of N particles?

This is where probability (again) enters into the story. In this context, the Born Rule says that the wave function's amplitude at a point (or better: the integral of the amplitude over some region) of configuration space determines the probability of finding the positions of the N particles to be such as to fall within that region, if we were to measure the positions of the particles. A wave function that is sharply peaked near a certain point of configuration space—that is, with amplitude near-zero everywhere except in a small region centered on that point—represents that if we measure the positions of the particles we will find them to have those specific values ($\pm\varepsilon$), with probability near 1. But a wave function that is smeared out broadly, or has many distinct lumps or bumps of non-trivial amplitude in different regions, represents our set of N particles as not having *any* definite positions, at the moment, and merely having certain probabilities of *being found* in certain places, if we go looking for them. The probabilities are, again, given by the Born Rule, which in the current context (roughly) says that the probability of finding the particles in certain locations is the square of the amplitude of the wave function for these locations. In other words, such a wave function represents the particles as being in a superposition state, a superposition of multiple possible precise location-states. Figure 7.5 illustrates this idea for the simple case of a two-particle system, with only one spatial dimension represented for each particle.

Now that we have seen how QM treats position, we are ready to raise our final concern about the Lewis-style approach to Humean chance in QM.

Regarding the wave function, we might wonder how such a thing, if taken as a physically real element of our world, fits into the HM. If we stick to Lewis' line concerning the nature of the HM, it seems

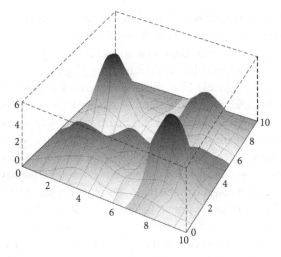

Figure 7.5. Wave function of two particles in one spatial dimension, in an infinite potential well (i.e., a box).

that there is a problem: the wave function cannot be defined in 3-d space, and straightforward attempts to boil its content down into something that *does* fit into 3-d space inevitably lose vital information encoded in the full wave function.[16] But redefining the HM to fit QM is also very problematic. Should it be configuration space that is the basic backdrop for the HM? If so, should it be the configuration space of position, or momentum? If instead we choose an infinite dimensional Hilbert space as the backdrop for the HM, then the universe's quantum state becomes a mere vector or ray singled out in that space;

---

[16] The fact that one of standard QM's key elements, the wave function, is defined in a high-dimensional space of some sort is a problem for more than just the Lewisian-Humean approach to chance; it is an issue that all those who would like to be realists about QM must confront. One approach to the concern is to try to use the fact (if it is one) that QM is only an effective, non-relativistic-limit approximation to some deeper, relativistic quantum field theory (which is set in ordinary spacetime) as a way to bring the wave function back down to reality. Ways of recovering the wave function from quantum field theory are explored in (Wallace & Timpson, 2010) and (Myrvold, 2015). Given the limitations and interpretive difficulties of quantum field theories themselves, we feel that this way to resolve the ontological issue is promising but far from being clearly workable.

what kind of a mosaic is that, and how can we discern patterns in it that correspond to familiar scientific regularities?

We don't here want to claim that no successful responses to these questions can be found, but we do want to highlight again the advantage that our pragmatic HOC has when it comes to capturing QM chance rules, and particularly those of standard QM. Our inclusive, multi-level understanding of the HM makes it easy to locate precisely the pattern of successful quantum experimental results that did, in actual history, lead to the acceptance of QM as a pattern ripe for systematization, by standard QM understood as an instrumental recipe for making probabilistic predictions. No matter what the real fundamental level of our universe is like, the meso- and macroscopic patterns of events in 4-d spacetime that we are familiar with must turn out to be *real*—to supervene in the right way on that underlying fundamental level—and, hence, to be grist for the pragmatic Best System mill. HOC is, in short, the best theory (among all extant reductive theories of chance) for capturing the probabilities of quantum theory in an elegant fashion.

7.4. Summing Up

Probabilities are found in physics in both deterministic and indeterministic settings. Either variety can be accommodated by HOC in natural and straightforward ways. In this chapter we looked at two of the most paradigmatic physical theories in which objective probabilities play a central role: classical statistical mechanics, and standard nonrelativistic quantum mechanics. In the former, we found that the Best System for a world in which SM is a useful theory could incorporate Humean chances in two ways. First, the Humean Mosaic might be such that the pattern of transitions from one thermodynamic macro-state to another give rise directly to Humean chances for such macro-state transitions. Second, more directly, the pattern of actual evolutions of systems well-described by SM—that is, the patterns of how the states of such systems, represented by points in phase space, begin and evolve over time through various macro-states and eventually to equilibrium—could be such as to have a simple and elegant systematization in the combination of classical mechanics + the Past Hypothesis Proportionality Postulate. In either or both of these ways (if sufficiently compatible), the chances of SM can plausibly turn out to be proper Humean chances.

When it comes to quantum probabilities, we found that the concordance between HOC and the theory is practically a marriage made in heaven: what the theory gives us is, on the face of it, a recipe for calculating probabilities for experimental observations. Since that recipe is enormously well confirmed by actual experiments, we have overwhelming reason to think that the Best System of Humean chances for our world will include that recipe in its compendium of rules. If we ignore the measurement problem and move beyond experiments out "into the wild," and think of QM as prescribing probabilities for such things as decays of radioactive atoms, scattering cross sections for particles moving through various media (e.g., x-rays through the atmosphere, neutrinos passing through the Earth, electrons tunneling past a potential barrier, etc.), we again have overwhelming reason to think that observable events in the HM for our world concord with QM, and so the case for QM being part of the Best System becomes even more compelling. This is a fact that will not change, no matter whether QM is replaced by an ontologically clearer successor theory at some future date or not, and no matter whether that theory is deterministic or not. HOC is in this sense ideally suited to be the right philosophical account of the probabilities given to us by quantum theories.

8

Chance and Causation

8.1. A Reductive Relation?

Analytic philosophers are wont to look for ways to reduce one concept, notion, or category to another or others; and this book is of course a good example of this tendency. When it comes to objective chance, a notion that springs readily to mind as potentially related in some reductive way is one that we have said little about so far: causation. This chapter will explore the connections between chance and causation, as I see them. No new theory of causation will be put onto the market, but the hope is that the points to be developed here will be of some service to the community of philosophers who work on causation and kindred concepts.

There are two ways that a reductive connection between chance and causation could go: reducing (in part at least) chance-facts to certain kinds of causal facts; and reducing (in part at least) causal facts to certain kinds of chance-facts. The former type of connection can be read into the primitive propensity view discussed in chapter 1.[1] A chance propensity can be thought of as constituting a partial, "non-sure fire" cause, where the causal power, or degree of "sure-fireness" (so to speak) is supposed to captured by the probability value of the chance propensity. I argued in chapter 1 that this way of understanding objective chances renders them mysterious and literally incomprehensible—we don't understand in any clear way what the meaning of a specific, numerical chance claim is supposed to be. The reader may or may not have been convinced by those arguments, but I have nothing to add to them here, so we will set aside this sort of chance-cause relationship.

---

[1] This is not to say that every advocate of the propensity view would endorse this way of speaking.

In the other direction, some philosophers have been tempted to offer one or another variant of a probabilistic account of causation; the rough idea of all of them is that for C to be a cause of E is for C's presence to *increase the probability of E*. There are variants in which C and E are understood as event types, and the kind of relation being analyzed is generic or "type" causation: the stereotypical example is the fact that smoking cigarettes causes lung cancer. And there are variants in which C and E are specific facts or event and the relation is that of singular causation: the stereotypical example is Susie's throwing of the big rock that caused the window to shatter. In both variants the core idea is that for C to be a cause of E is for it to be a *chance-raiser* (or probability-raiser) of E.

Suppes (1970) presented a probabilistic account of generic causation that was distinctly empiricist in character. The probabilities of Suppes' theory were to be understood in good empiricist fashion as *statistical* probabilities, i.e., frequencies (in the right sorts of reference classes). As happens with most reductive analyses, the Suppes theory quickly fell beneath an onslaught of problems and counterexamples.[2] One response in the literature to try to save the basic idea of a probabilistic account of causation was to move away from actual statistical frequencies and to a more metaphysically loaded notion of objective probability. Unlike mere statistics, the *real* probabilities will not give us the intuitively wrong causal conclusion because of simple "bad luck" or too-small reference classes. But if the "real" probabilities are understood along the lines of primitive propensity-type chances, they become epistemically inaccessible and subject to the criticisms of chapter 1 already mentioned.

A somewhat different tack was taken by Lewis (1986b) and later by Schaffer (2001): to focus on token causation and marry the probability-raising idea to Lewis' counterfactual-based analysis of causation. The result is an approach that says, roughly: c was at least a partial cause of e if c's presence raised the chance of e's occurrence compared to what it would have been without c. Lewis' and Schaffer's theories also

---

[2] The essence of the problem with Suppes' theory is made extremely clear in Cartwright's classic paper/chapter "Causal laws and effective strategies" (1979/1983). The critique is extended further in my "Humean effective strategies" (Carl Hoefer, 2005).

fell beneath the ravening causality-literature counterexample hordes. Here, the problem was that, for singular cause-and-effect events, there can be cases where the (or *an*) actual cause *lowered* the probability of the effect (but it happened anyway), and cases where an event *d* raises the probability of the effect *e*, the effect happens, but intuitively *d* was not a cause of *e* (because the causal pathway from *d* to *e* was blocked or preempted). We will see examples illustrating these problems shortly.

There is a further general problem for the whole idea of reducing causal facts to objective probability facts that is worth remarking: there may not be enough probability facts out there in the world to do the job! As we saw in previous chapters, this is certainly a concern if we understand probabilities as HOCs. For type-level causation, the Best System may simply not cover certain types of events that nevertheless are causes of certain effect event-types (and the presence of which, intuitively speaking, does "raise the probability" of the effect event-type—a tension that will be resolved later in this chapter). For singular causation, which is ubiquitous (all around us all the time), it is clear that the HOCs can provide the required chances only if the Best System yields physical probabilities for every possible evolution posterior to any arbitrary given starting conditions (as *might* be the case if something not too dissimilar to quantum mechanics figures in the world's Best System). Even if one is confident that such probabilities exist in our world, one might want to think twice before analyzing causation in terms of facts so epistemically inaccessible to finite agents such as ourselves. And in any case, the counterexamples mentioned one paragraph back still threaten the whole project.

So attempts to reductively analyze causation to probability seem to be doomed, and for a simple reason: whatever causation *is* (if anything), it is not *simply* the raising of effect probability. As Anscombe concluded in her celebrated work, "Causality and Determination," it has something to do with *bringing about* (as in "*c* brought about *e*," or "*c* made *e* happen"). And while probability-raising and bringing about often go together, they can and sometimes do come apart.

In the next section we will consider a paradigmatic example of a cause that (intuitively) raises the chances of an effect, which effect does happen; and some twists on the example that illustrate some of the difficulties mentioned in the preceding. Nevertheless, there do

seem to be connections between probability and the cause-effect relation, and simply noting that they often go together does not clarify
the connections sufficiently. So in the final section I will propose and
defend a principle that articulates the strongest assertable connection
between causation and probability. As we will see, it is not chance,
but rather *subjective probability* or credence, that is involved in the
connection.

8.2. Bob's Great Presentation

At the ad firm where he works downtown, Bob has to make a presentation of a new ad campaign he has created, to his boss and the
prospective client (and a few colleagues of Bob who might work on
the campaign if it goes ahead). Bob does his presentation, and hits the
ball out of the park; it's all smiles and back-slapping when the meeting
ends. On the way back to their cubicles, Bob's colleague Lisa remarks,
"Great presentation, Bob, you've really upped your chances of getting
that promotion!" Two weeks later, Bob does get his promotion.

Let's suppose that Bob's great presentation is indeed one of the
causes of his later promotion, because it really impressed the boss and
changed her view about Bob's talents and the right place for him in the
company hierarchy. Now, was Lisa right? Did Bob's giving the presentation raise the chance of his getting the promotion?

First, let's note that we are talking here about singular causation.
The Best System will probably not contain a generic chance for the
outcome *get a promotion* given the setup condition(s) *give a great presentation at work*. There are too many reference class options available,
both on the setup side and the outcome side. What features make a
presentation *great*? What range of companies form the supervenience
base for the generic chance: ad firms in the United States? Ad firms
in the Northeastern United States? Or just in New York? Or ad firms
plus a wide range of other types of corporations and firms? How long
after the presentation may the promotion occur? Does a standard
good-performance annual pay rise count as a promotion? During
which years does the chance remain stable? And so forth. There's no
reason to expect that the Best System trade-off between simplicity and

strength/utility will result in any of these potential candidates making it into the System.

So if Lisa's allusion to Bob's chances is to be taken literally under HOC, we have to be talking about micro-derived chances, which—in some sense at least (see chapter 6)—may be taken to cover all sorts of singular as well as generic events. Now let's come back to the question: was Lisa right, did his great presentation up Bob's chance of getting the promotion he desired? The correct answer is: *maybe yes, maybe no; and the yes/no answer depends on a myriad of other factors, inside and outside the office.* Bob's presentation may well have lowered his chances of getting a promotion—because, say, the great presentation event raised—dramatically—the chance that Bob would go out for celebratory drinks that evening with his colleagues, and thus the chance that Bob would be heading home on the 20:04 train instead of his usual 18:34 train; and the 20:04 train has a terrorist bomb planted on it that will cause the deaths of most passengers. In this unfortunate scenario, Bob's presentation lowers his chance of getting his promotion by lowering (dramatically) his chances of being alive the next day; dead men get no promotions. Now, as it happened, Bob did go out and celebrate, but (quite improbably) stayed out for one more pint than he intended and missed the 20:04 train, saving his life and thus the possibility of eventually being promoted. But that was a lucky, *improbable* turn of events, from the perspective of when Lisa uttered her claim about Bob's chances. What Bob's great presentation had done, as of the moment Lisa made her claim, was lower Bob's chances of getting that promotion.[3]

---

[3] Here and below I'm invoking commonsense causes and effects (impressing the boss, getting blown up, etc.) in the course of making understandable how the micro-derived chance of Bob getting promoted could have been lower after making his presentation. In so doing, I'm offering a causal/explicatory gloss on what are calculations strictly in the province of a Laplace demon, as we noted in chapter 5. Think of it as what the demon would tell us by way of simplification if we asked him how it could be the case that Bob's chance of promotion was lower just after his presentation than it had been before.

But we should keep in mind that such a commonsense gloss need not be available, for this kind of intuitively unexpected probability shift to occur. It could happen just because of a myriad of facts about the micro-motions of particles, none of which can be easily summarized in a humanly accessible causal narrative. "But that's incredibly

One might suppose that once he fails to catch the 20:04 train, his micro-based chance of being promoted zooms up again, and to a level higher than it was before his presentation. Again, maybe yes, maybe no; perhaps Bob's presentation itself probabilified other events (e.g., machinations by jealous colleagues) whose impacts on his promotion chances would be negative enough to make his promotion chances remain lower than they were pre-presentation, right up to the day before the boss decided to promote him. If the boss nevertheless gives Bob the promotion, and does so (in her own mind!) partly *because* of his great presentation, we should judge that the presentation was a partial cause of the promotion. The lesson seems to be simple: causation and chance-raising are not the same thing.

What about looking instead at counterfactual chance-raising? Well, so far we have been comparing the probability of promotion just prior to the presentation with the probability just afterward. But in exactly the same scenario, plausibly the probability of promotion is also lower than *what it would have been* in the nearest non-presentation world, on any plausible way of determining that. As long as Bob stood a reasonable chance of getting the promotion anyway (say 30%) and the great presentation bumps his probability of an early demise up to (say) 80%, the counterexample still works.

The reader may still feel uneasy about the conclusion being drawn from these counterexamples, and may be tempted to express the unease like this: "Granted, given the contrived circumstances you craft, the probability of promotion goes down instead of up. But surely Bob's great presentation raised the chance of his promotion compared to what it would have been, *in each of the counterfactual circumstances where we hold fixed all causally relevant facts intermediate between the presentation and the promotion?* That is, compare the counterfactual probability of promotion with and without the presentation, but holding fixed +/- each variable that is also a cause or preventer of his promotion (being on the 20:04 train or not, machinations of his jealous colleagues, etc.). It remains plausible that once we hold fixed

unlikely!" you may wish to protest. True (perhaps), but so what? A counterexample to an analysis is still a counterexample, even if its occurrence is infrequent or unlikely.

any other potential interfering (or helping!) factors, Bob's chances *with* giving a great presentation are higher than his chances *without* doing so.

This is just adapting Nancy Cartwright's famous (1979) principle *CC* to counterfactual chance-raising. This may, then, get things right for the range of Bob cases so far discussed; but (a) it ruins any possible *analysis* of causation (since "causes" appears on the right-hand side), and (b) it may be possible to find awkward counterexamples to this counterfactual *CC* condition too (as is suggested by a minimally realistic induction on attempts since 1979 to make both counterfactual analyses and *CC*-like principles work), and (c) it creates a new problem of non-existence of probabilities. From the *actual* microstate of things, you may be confident that there exists Ch(Promotion | great pres, celeb drinks, catch 8pm terror train, . . .)—where all these things actually have fixed chances of happening or not, given the microstate just after the presentation. But what determines Ch(Promotion| ¬Presentation, . . .)? What would the microstate have been, if Bob didn't give the presentation? What's the closest world, one where he was never asked to do it, or where he got sick on the morning of the day? Familiar problems from the literature on counterfactual analyses of causation arise here, but they are arguably worse: there are an infinity of ways to flesh out the microstate of a counterfactual world in which Bob gives no presentation, and there's no way to pick out the "right" one or right set of them, and in case we choose a set, no "right" way to average over them to extract a single numerical conclusion.[4] But I will not pursue the arguments any further since point (a) is sufficient to show that we are not being offered here a candidate reductive analysis of causation in any case.

In light of the preceding, the reader may agree that there are problems facing any attempt to reductively define causation in terms of probability-raising, but still feel that there's something unfair and nit-picky about the problems being raised. Whatever the merits of the preceding counterexamples *qua* counterexamples to particular

---

[4] The problem alluded to here is analogous to, if not identical to, the problem of choosing the "right" measure over a continuous phase space in statistical mechanics; see chapter 7.

philosophical theories, the reader should still feel that what Lisa told Bob was *right, damn it*: his promotion was more likely after giving such a great presentation! But how can Lisa be right if the terror-bomb train was looming in Bob's near future? Well, she didn't know about the terrorist bomb, and anyway such things are so unusual that we rightly ignore them in coming to judgments about the likelihood of future events; Lisa may not have been right, literally speaking, about objective probability; but she was surely being epistemically rational.

The point can be re-phrased like this: "Since such awkward circumstances are rare, and in any case normally beyond our knowledge, surely we are right to ignore their possibility, and increase our *credence* in the effect's occurrence, when we are told that a(n intuitive) cause has been introduced." Exactly! And this brings us to the *correct* relation between causation and probability, which I will baptize as a Principle: CCP, the Cause-Credence Principle. The connection between causation and probability is epistemic, rather than ontic.

8.3. The Cause-Credence Principle

Lisa may have been making a claim about the objective chance of Bob's getting a promotion, if her worldview includes such things. But in general, when people say things about the chances of this or that going up or down, what they almost certainly *are* expressing is something about their own degree of belief in the future outcome at issue. (If they are in fact intending to say something about a change in objective chances, then the change in personal credence follows immediately from PP, which most agents' credences satisfy.) We can take it that, if Lisa was disposed to think a bet in favor of Bob's getting promoted would only be fair at relatively long odds, she now would consider such a bet fair at much lower odds.

This, I submit, is the right sense in which we should understand causation and probability-raising to be connected: *ceteris paribus*, when a rational agent learns that a cause for some possible effect has been introduced, her credence in the occurrence of the effect (if she has any) should go up; and if she learns that a preventer of some possible effect has been introduced, her credence (if she has any) should go down.

In this section I'll try to articulate this more precisely, and discuss what has to go into the crucial *ceteris paribus* (c.p.) clause. First the Principle:

> **Cause-Credence Principle (rough version)**: *if you don't know whether a certain effect is going to be produced (prevented) or not, and you learn that a cause (preventer) of the effect has been introduced,* ceteris paribus *you should increase (decrease) your subjective degree of belief in the occurrence of the effect, on the relevant occasion.*

As noted, the c.p. clause is crucial here. It needs to be explicated, because c.p. clauses without explication render a proposition empty or tautologous; and it must be fleshed out in a way that does not turn CCP into an explicit tautology.[5] Our explication will have to be crafted in a way that evades certain sorts of clear *prima facie* counterexamples to the principle, without becoming a laundry list of fixes and patches.

The first thing to understand as intended in the c.p. clause is an exclusion of agents who have no current credence (whether sharp or interval-valued or modeled by a family of functions, or what have you) in the proposition that the effect e will occur. In Box 1.1 of chapter 1, I argued that it is no requirement of rationality that agents have credences in all propositions they can understand. If an agent has no credence level at all in $<e>$, we can hardly require that she *increase* that level when she learns that a cause of e has been introduced.[6]

---

[5] Consider, for example, "All A's are B's, ceteris paribus." If nothing is said about the c.p. clause, then we learn essentially nothing from this claim. It potentially is compatible with anything from no A being a B to all A's being B's. Of course, the point of making an assertion like this with a c.p. clause is usually to indicate to the reader that most A's are indeed B's, except when certain exception-conditions obtain. But if we make this—with no further explication of those exception-conditions—the content of the c.p. clause, we end up with an uninformative tautology: "All A's are B's except in the circumstances in which they are not."

[6] What if the agent learns that a *deterministic* cause of e has been introduced; surely her credence should then pop into existence, and indeed be (nearly) 1? I don't think this can be seen as a requirement of rationality, unless the agent has an understanding of "deterministic cause" that is strong enough to *logically* imply e (not merely nomologically, or with-chance-1, or . . .). If she does have such a concept, then the ordinary strictures of subjective probability take over: it is a theorem that if an agent assigns

To understand a second element needed, let's look at a counterexample to CCP from Alan Gibbard:[7] Imagine you are flying to the beautiful tropical island Bella Isola for vacation. The pilot gets on the intercom and announces: "Good news! We've just been informed that the Bella Isola authorities recently sprayed oil in the swamps, so there will be fewer mosquitoes around and less chance of anyone catching malaria!" Your credence in M: <I catch malaria while on vacation>, which had been quite low (say 0.1%, since you had no idea it was an issue on Bella Isola) now zooms up to something significant (let's say an alarming 20%); but according to CCP your credence should have gone down, not up.

What has happened here, clearly, is that your credences have undergone a shift with two distinct components. First, you learned (by implication from the pilot's remarks) that contracting malaria is something to be worried about on Bella Isola. You didn't know this before, but if you had, your credence in M would have been much higher than 0.1%. Second, you learn that the swamps have been sprayed, which has some effect in reducing the frequency of malaria cases; so your credence in M ends up being lower than it would had you *only* learned that malaria was an issue on Bella Isola. When we distinguish these two components, the counterexample looks less problematic.

A tempting way to address the problem is simply to add the word "only" in the right place:

CCPonly: *if you don't know whether a certain effect is going to be produced (prevented) or not, and you learn **only** that a cause (preventer) of the effect has been introduced,* ceteris paribus *you should increase (decrease) your subjective degree of belief in the occurrence of the effect, on the relevant occasion.*

Since the pilot's announcement induces you to learn more than just of the introduction of a malaria-preventer, CCP does not apply to you, and your big bump upward in credence in M is no counterexample.

---

probability 1 to <c> and as a matter of logic <c> ⇒ <e>, then the agent must assign probability 1 to <e>. So CCP need not be involved in this case.

[7] Prof. Gibbard offered the counterexample in a graduate seminar I taught at U. Michigan in 2008.

Unfortunately, the insertion of "only" does not seem to handle other sorts of apparent counterexamples. Suppose you already knew that a cause c_1 had been introduced, making e more likely, but now you learn (only) that cause c_2 has also been introduced. But you already knew that c_2 is a less effective promoter of e than c_1, and moreover that c_2 interferes greatly with the effectiveness of c_1. In such circumstances you might reasonably find your credence in <e> lower, after learning of the introduction of c_2.[8]

The problem seems to be not so much a matter of what, precisely, one *learns*, what new information one gets, but rather whether that new information interacts with previous knowledge in a way that affects one's credence in <e> in some way independent of the "normal route" by which learning of a cause (preventer) being introduced leads to raising (lowering) credence in <e>. (What this "normal route" amounts to will be discussed shortly.) So we need to change CCP, or flesh out the c.p. clause, in a way that de-activates CCP when those non-normal routes of impact are present in the agent.

One way to do this is to follow the model of *admissibility*'s definition for PP. Recall that information E is admissible for a given application of PP (to a proposition A, for a rational agent $Cr(\_\_)$, with background knowledge K, *iff* the impact of E on reasonable credence concerning A, if any, comes only by way of impact on credence concerning the objective chance of A (and is therefore canceled by X, which stipulates A's chance outright).

A similar admissibility clause for CPP might go like this: *if learning of the introduction of the cause* c *(preventer* p*) does not, when added to* K, *convey to the agent information relevant to the likelihood of effect* e *where the relevance comes by way of something other than judgment concerning the likelihood of the normal causal paths or mechanisms* c (p) → e (¬e) *being successful on the relevant occasion.*

---

[8] This is an example of interference or preemption. We also need to rule out cases where the agent knows there is an equilibrating mechanism in play. Example: a hike in interest rates is *ceteris paribus* a preventer of borrowing (e.g., MasterCard spending), but if I know that the current CEO of MasterCard only raises rates while simultaneously blanketing the airwaves with new advertisements so as to not lose revenue, I have inadmissible information vis à vis application of CPP to this case.

The admissibility criterion here is arguably even more awkward and engineered-looking than PP's. And it may still require further refinement. But it seems to be roughly what is required, if something like CCP is to be maintained at all. Putting the pieces together we get our definitive version of CCP:

> **CCP: IF** *an agent does not know whether a certain effect* e *is going to be produced (prevented) or not, but*
> i.   *has a certain subjective credence in* e's *occurrence, and*
> ii.   *learns that a cause* c *(preventer* p*) of* e *has been introduced, and*
> iii.   *learning of the introduction of the cause* c *(preventer* p*) does not, when added to K, convey to the agent information subjectively relevant to the probability of effect* e *where the relevance comes by way of something other than judgment concerning the probability of the normal causal paths or mechanisms* c (p) → e (¬e) *being successful on the relevant occasion;*
> **THEN** *the agent should increase (decrease) their credence in the occurrence of* e, *on the relevant occasion.*

CCP is quite a mouthful, lacking the simplicity and clarity of PP. But I believe that it captures a true, general connection between causation and probability.

Notice that CCP is not even a partial analysis of causation. It relies on a prior understanding of causation, and tells us what the relationship of causation and probability ought to be: namely, a strongly hedged, *ceteris paribus* relationship of *impact*, knowledge of causes impacting (causally!) agents' reasonable degrees of belief. What is that prior understanding of causation? What is intended is the quite general Anscombian understanding: causation is *bringing about* or *giving rise to*. The admissibility clause, however, alludes to a "causal path or mechanism." I think this does not substantially restrict the range of types of causation covered by CCP, because *mechanism* is intended here in a very broad sense. A cause is something that, when introduced into a situation that did not have it (and would not have had it, absent the introduction), opens up the possibility of operation of a mechanism that brings about *e*. By "mechanism" I mean a chain of events with the right sort of connections, whether those connections are thought of as *reliable, probabilistic,* or even *deterministic,* whose operation may result in *e.* The chain may have no intermediate links, e.g., if the cause *c* is the presence (rather than absence) of a radium atom and the effect *e* is the subsequent decay event leaving an alpha particle and a polonium atom. The chain, or certain links in it, may also only operate with a certain objective probability, as the radium decay example also illustrates (if we specify that the effect *e* is, say, "decay event within the next hour"). Or the chain, or certain links, may only operate with some degree of reliability—which could even be low, subjectively speaking, e.g., when one takes an aspirin hoping to relieve a headache, or when doing chemotherapy hoping to cause remission of an advanced-stage cancer. No objective chances need be involved in such chains, or any of their links, in order for the operation of the chain from *c* to *e* to be a possibility; and it is that *possibility* that grounds the rational force of CCP.

With this broad Anscombian understanding of a causal path or mechanism in mind, the rationality of CCP should be fairly evident. To learn that a cause has been introduced, and to satisfy the admissibility clause, is to learn exactly one thing relevant to whether or not *e* will occur: that there is now at least *one* causal path in existence that may lead to the occurrence of *e*, which one did not know was in place previously. In such circumstances, *if* one had a certain subjective credence level in <*e*>, that credence level should be increased after learning <*c*>. If *c* is an ordinary cause of *e* (and not, at the same time, a preventer of *e*), how could the credence stay the same?[9] That would be equivalent to the agent's being certain that *c* will *not* end up leading to the occurrence of *e*; but if she satisfies the admissibility clause (and understands the notion of "*cause*" in play here), the agent could not reasonably have such certainty. Therefore the agent would be irrational if she maintained the same level of credence as before

---

[9] Causes that are at the same time preventers of a certain effect raise complications that I do not want to discuss in detail. Admissibility clause (iii) of CCP is to be understood as violated if the agent knows that a certain *c* is both a promoter and preventer of *e*. CCP could perhaps be extended to cover causes that are also preventers, in cases where the agent believes that *c* is clearly a stronger cause than preventer (or vice versa), but it is not something I wish to attempt here.

learning of <*c*>—which shows that CCP is, plausibly at least, a principle of rationality.

CCP may seem disappointingly near-tautologous. But in fact, it seems to me that most principles of rationality should seem near-tautologous, or outright tautologous, given that respecting the tenets of logic is the essence of rationality. PP itself seems quite self-evident at first sight, until one stops to reflect, "What exactly is an *objective chance*? And why should learning that such a thing has a certain value affect *my* credences?" When the answers to these questions are fleshed out by HOC, as we saw in chapter 4, the result is a satisfying, if hardly immediate or obvious, deductive provability of PP. With CCP what needs to be fleshed out is what is meant by *cause (/preventer)*. Some remarks about that issue have already been made in the preceding; in the rest of this chapter we will expand on those remarks and compare the roughly "mechanist" understanding of causation in play with other conceptions from the philosophy of causation literature.

8.4. More on Causation

The preceding talk of "bringing about," "reliable connections," "operation of a mechanism," etc., is all causally charged talk; neither the previous brief description of what a "mechanistic" approach to understanding causality amounts to, nor any other I have seen, tries to *reductively define* causality in terms of something else that is clearly and distinctly non-causal. This is not a problem, because we are not trying to analyze causality here, just understand its connection to probability. Still, more can be said about why it is natural to think in a mechanist way about causation, and why that way of thinking is more useful than certain other prominent approaches.

First let's come back briefly to the approach of probabilistic theories of causation, i.e., theories that analyze causation as objective chance-raising (or counterfactual objective chance-raising). Notice that if causation were objective-chance-raising, then a special case of CCP would follow more or less directly from PP. If the agent has a credence in <*e*> that already was fixed by PP, then learning that a chance-raiser had come into play would seem to force the agent to increase her credence in some (perhaps vague or interval-valued) way, even if she does not know what the *new* objective chance of <*e*> happens to be. I will not try to demonstrate this consequence here, however, because we already have seen two important objections to analyzing causation in terms of chance-raising: causes that are in fact chance-reducers (viz., Bob's great presentation) and the possible lack of enough objective chances to ground all the causal facts we believe in (as discussed in section 8.1).

Earlier I described a mechanism or causal path as a chain of events with the right sort of connections, whether those connections are thought of as *reliable, probabilistic,* or even *deterministic,* whose operation may result in *e*. This may seem like a roundabout way of saying what the probabilistic theory defender is trying to say: a cause is the kind of thing that makes the occurrence of the effect at least more probable, and perhaps certain (if all links in the chain are infallible deterministic connections). And this is right, but my point is that "more probable" should be understood in a subjective/credence sense, rather than in the sense of objective probability. The not-enough-chances objection is crucial here: if there is, in fact, no determinate objective chance of *e* occurring, either before or after the introduction of *c*, then when we say that *c*'s introduction made *e*'s occurrence more probable, we can only mean that we are justified in *acting as though e is more probable,* i.e., raising our credence in *e*. And that is just what CCP aims to capture. But it does not show, in any way, that *what causation is,* is objective chance-raising.[10]

Why do I emphasize mechanism rather than (say) counterfactual conditionals (à la Lewis), invariance (à la Woodward), or INUS conditions (à la Mackie)? Each of these accounts captures important aspects of our thinking about causation, to be sure; but it seems to me

---

[10] I take it as obvious that an account of causation that analyzes causation as *subjective probability raising* is a non-starter, making the presence or absence of causality be a personal and subjective fact that can vary from agent to agent. If we instead consider a possible account of causation as *rational-agent probability raising* we can avoid the subjectivity and agent-relativity. Such an account would have to be fleshed out considerably in order to have even minimal plausibility, and the fleshing out would probably take things in the direction of the *epistemic* theory of causation recently advocated by Russo and Williamson (e.g., 2011). Their approach will be briefly discussed in the following.

that the notion of mechanism, broadly understood, together with the notion of *laws of nature*, lets us understand why and how these accounts do well, insofar as they do, and why they go wrong where they do. In this sense, these notions (mechanism or causal path; laws of nature) seem to me to be more fundamental, which justifies that CCP invokes the former, rather than some other conception of causation.

Consider a standard cause-effect relationship and how Lewis' counterfactual approach handles it: Suzy throws a heavy rock at a glass window (c), shattering it (e). Lewis' theory delivers the verdict that c caused e if the nearest c-world is an e-world (actuality; check) and the nearest $\neg c$ world is a $\neg e$ world. Now, how do we decide whether the counterfactual $\neg c \; \Box\!\!\rightarrow \neg e$ is true? It seems obvious: take away the thrown heavy rock, and of course the window does not shatter! But c is a quite specific event: a very specific, singular rock thrown in a specific manner at a specific time in a specific direction and angle, etc. Vary any of these features of Suzy's rock-throwing and what we get is, arguably, a distinct event, not c. Doesn't that mean that the *nearest* $\neg c$ world is one where only the angle, or the force, or the timing of Suzy's throw is different, but all else as much like actuality as can be? But that is not what Lewis' theory needs, of course, because reading the counterfactuals in this way yields the incorrect conclusion. In those very close $\neg c$ worlds the window still shatters, which would entail that according to Lewis' theory Suzy's throwing the rock did *not* cause the window to shatter.

Lewis (1973) was not tripped up so easily, of course.[11] He insisted that the identity conditions for an event like c are not so fine-grained; in a world where Suzy throws the same rock but 1° higher or lower, we have the same event c occurring. But what grounds such judgments? Why is it the same event if Suzy throws 1° or 2° to the left of how she actually throws, but not if it's 30°? For that matter, if Suzy is throwing at a large window that wraps around her in a 180° arc, then is an event where she throws 30° to the left (still well within the arc of the window) still c, in that case?

---

[11] In the later (D. Lewis, 2000) theory of causation in "Causation as Influence," Lewis takes a different path to handling this issue of what constitutes $\neg c$, and offers a substantially different account of causation there. I will not discuss Lewis' later approach.

It seems to me that what guides our intuitions in thinking about what should be considered a ¬c world is the notion of mechanism: a proper ¬c world should be a world where the window-shattering mechanism [girl-throws-heavy-object-directly-at] is absent. And thus, a world that differs from actuality in even substantial ways (Suzy throws 30° to the left of her actual trajectory; throws a different rock, or a metal paperweight; throws at twice the velocity; etc.) does not count as a ¬c world, as long as that mechanism is still present and activated.[12]

Consider also the sorts of cases that did trip up Lewis' 1973 theory, namely preemption cases. In actuality, Suzy threw her rock and it broke the window, but Billy also threw a heavy rock, equally hard, just a second later than Suzy (and before her rock hit the window). In this scenario it is clear that (a) Suzy's rock-throw does shatter the window, and (b) in the nearest ¬c—world, e still occurs, because of Billy's throw. So Lewis' theory delivers the intuitively wrong verdict: that Suzy's throwing the rock did not cause the window's shattering. One may be tempted to say that in the nearest world where Suzy does no throwing, the event of the window shattering is not to be considered as the same event e. But we already decided that e could happen a second earlier or later and still be e, so it is not clear on what grounds we can assert that the window shattering is not e in this scenario.

There are ways to try to patch up the theory to avoid preemption counterexamples, though new counterexamples always seem to arise and the project eventually develops an epicyclic feel. What I wish to highlight is that our intuitions about all cases—where the theory works, and where counterexamples threaten—are driven by our assessment of *what brings about what*, and by *what route/path/mechanism*.

The same observations can be made regarding Mackie's early INUS-condition account of causation (1965). In that account, a type-level theory, an event type c is a cause of event type e if it is an Insufficient but Necessary part of an Unnecessary but Sufficient condition for the

---

[12] In order for the counterfactual theory to deliver the right verdict, these different possibilities cannot be ¬c worlds because it is pretty clear that some of them, at least, are closer to actuality than a world in which Suzy simply doesn't throw anything.

occurrence of an e event. In the first form of the theory, Mackie gave an empiricist gloss on the meaning of "necessary" and "sufficient" here, roughly: A is necessary for B *iff* there are no cases of B-events (anywhere, anywhen in the world) that are not accompanied by an A-event. And D is sufficient for G *iff* there is no case (anywhere, anywhen in our world) where D occurs and is not accompanied by G.[13] The analysis faces potential counterexamples of at least two types:[14] (1) there may be factors D which simply *happen to* always occur jointly with certain others in making up an US condition, but which are intuitively not causally related to the effect; (2) there may be certain pairs of joint effects of a common cause such that one of the two effects satisfies the definition of an INUS condition with respect to the other, but intuitively there is no causal link or path between them (only a shared cause in their past). As with the counterfactual theory, a simple diagnosis can be given of the source of problems: the theory is defining the causal relation in terms of what is (intuitively, on reflection) merely a *symptom* of causal connection (or lack of same)—and not an infallible symptom. If A and B (perhaps together with a list of other factors) always go together in our world, it may be that there is a causal or nomological connection between them; but then again, it may be a mere happenstance regularity.

In most of the cases of counterexamples to theories of causation that you will find in the philosophy literature, intuitions are fairly strong and universally (or nearly) shared that the theory gets it wrong; and usually that is so because one can easily discern "what's really going on" in the example, i.e., the real causal paths and mechanisms at work. Conversely, in many cases where the causal path is unclear or multiple

[13] I am offering here a simplified sketch of Mackie's 1965 account, but the simplifications do not affect the points I wish to make. Here I use "accompanied by" rather than temporal phrases such as "preceded by" or "followed by," simply because it was common among empiricist regularity theorists of causation to not wish to presuppose that causes always precede their effects (see, e.g., Mackie, 1966).

[14] The theory is also unable to handle the possibility of purely non-deterministic causation: causation where the cause may bring about, or fail to bring about, the effect, on different occasions, even when everything about the situations is held fixed. It is not obvious to all philosophers that any such cases exist, but many (e.g., Cartwright, Anscombe) have vigorously defended the likelihood of such cases in nature (Anscombe, 1971; Cartwright, 1989).

paths intersect, our intuitions about what is and isn't the cause are correspondingly muddied. When a 10-man firing squad executes a prisoner with 10 simultaneous and well-placed shots, whose shot caused the death? Did none of them individually cause it, but only the squad's action as a whole? A mechanist perspective upholds this plausible (if not overwhelmingly clear) viewpoint. Each soldier's shooting is, on its own, an example of an effective mechanism *type* for bringing about death, but each particular shooting may be held to not actually *bring about* the death on this occasion; but surely the firing of the whole squad is both an effective mechanism and *did* bring about the effect, on this occasion.

Or consider a less gruesome example: a motorist enters an intersection and hits another car, causing an annoying fender-bender. The driver was, it's true, looking down at his radio as he approached the intersection and didn't see the other car. He had, however, seen already that there was no stop sign at the intersection. There was no stop sign because, unbeknownst to him, on the previous night some teenage vandals had uprooted and stolen the stop sign that normally was there. Finally, the roads were also slick from rain, making our motorist's car skid when he slammed on the brakes, rather than coming to a stop in time to avoid collision. What caused the accident: The theft? The driver's inattention? The rain? The joint presence of all of them? Or is there in fact no cause worthy of the name here?

Each of the three types of analysis of causation in terms of something clearly different (probability, counterfactual dependence, regular co-occurrence) seems to suffer from the problem of mistaking a symptom or upshot of causal connection for causality itself. Noticing this fact, many philosophers urge us to give up on trying to analyze or define causation in terms of something else; just accept that it is a *primitive* notion that is indispensable to us, in both science and daily life.

Here's what I don't like about that: making causality into a primitive, rock-bottom notion means that we give up on saying anything *further* about why *c*'s bring about *e*'s, or about what that bringing-about consists in, in a given occasion. But that seems like giving up too quickly, because in fact in very many paradigmatic cases of causation, we *can* say a lot more about how and why *c* caused *e*, or *c*'s tend to cause *e*'s, and what the bringing-about or causing relation amounts to. We do that by (1) specifying the mechanism at work, and (2) grounding, as far as possible, the necessity of the individual links of the causal chain in *physical* necessity. Specifying the mechanism amounts to establishing the initial conditions, the setup, the particular circumstances (of both *c* and its environment, in general) that constitute the mechanism; if we dig down into the details of how the bits and parts of the mechanism push and pull each other to make things occur, we ideally find one of two things: either a lower-level mechanism, which in turn can be taken apart, studied, understood; or the bits and parts affecting each other in ways that can be seen to be direct consequences of physical laws (e.g., when a biological mechanism is analyzed down to the level where ion transport across a cell boundary is one link in a causal path). At that point, one has an answer to "Why does process *p* happen?" or "Why does *x* bring about *y*?"—an answer that does not advert to causation, mechanism, or any cognate notion. The answer is: because for *p* to not happen in these circumstances (or for *x* to not bring about *y*) would require violation of a law of nature. Physical causation gets its modal *oomph*, insofar as it has any, from the necessity of physical laws.[15] It may not be possible in all cases—may be not possible in *most* cases, even—to take apart causal mechanisms or path and break them down into bits in which each bit's action can be seen as underwritten by nomological necessity. But sometimes it *is* possible to do so, and for that reason I believe we should resist making the cause-effect relation into a metaphysical primitive.

Finally, let's come back now for a moment to CCP. To learn that a cause for effect *e* has been introduced is to learn that something has taken place which *can* lead via some path to *e*; or which even, if circumstances are very propitious, *must* lead to *e*, with the modal force of the *must* coming from the laws of nature. An epistemic possibility is opened up for the rational agent, concerning how the world may unfold; and if we have correctly designed our *ceteris paribus* clause, then if the agent's overall epistemic state satisfies the

[15] I write "physical causation" here rather than just "causation" to be cautious. Mental causation and agent causation, if such things exist, may get their "oomph" from something other than (or in addition to) nomological necessity.

c.p. clause, she would be irrational *not* to increase her subjective probability in the occurrence of *e*. By "cause" in CCP, I have indicated that we should understand "causal mechanism" in a very broad sense. The mechanism could be a simple one-stage process underwritten directly by physical law (e.g., when I create electromagnetic waves by waving a charged ball on the end of a stick back and forth in front of me). Or it could be an elaborate many-stage mechanical mechanism.

Personally, the more mechanism-like a causal mechanism, the better I like it. So my ideal for understanding the notion of "introduction of a causal mechanism" for an effect is a Rube Goldberg device; e.g., one of Wile E. Coyote's elaborate mechanisms for catching the Roadrunner. Once such a mechanism has been set up and put into action, as events unfold it may or may not lead to the relevant effect; and there may, but need not, be an objective chance of the effect happening; and the introduction of the mechanism may itself in some roundabout way make the effect *less* likely (e.g., if setting it up makes enough noise to warn away all Roadrunners).[16] But learning of the Rube Goldberg device's being put into action will, if we satisfy the CCP's ceteris paribus clause, give us *some* new reason for thinking that the effect may ensue, and *no* new reason to think it will not. And that is what CCP captures.

8.5. Summing Up

Objective chance cannot be analyzed as some kind of causation, nor can causation be analyzed in terms of something like increased objective chance. Nonetheless, we do commonly think that the presence of a cause increases the probability of an effect, and we speak in those terms in everyday life. By thinking through one specific example of such cause-probability talk, we came to see that the

---

[16] This would, if known, constitute inadmissible info for CPP's use here.

The reader may feel that my example is ill-chosen, since the Coyote *never* catches the Roadrunner. But this is actually not true; sometimes he does catch the Roadrunner, momentarily. What he *never* does is keep hold of him long enough to get a square meal.

real connection between causation and probability lies on the side of *credence* rather than the side of objective probability or chance. A cause-credence principle (CCP) seems to be defensible, but it is not something very central to understanding either causation or probability.

Concluding Remarks

This book has attempted to do something that is certainly out of fashion in academic philosophy today, and that is thought by many to be a hopeless task: to give a successful reductive analysis of a controversial, philosophically important notion. In my own view, the task has been accomplished, as cleanly and convincingly as one could hope for in a philosophical analysis. But I expect that a majority of readers will judge differently. So, in a last attempt to defend the virtues of my version of Humean objective chance (HOC), I will try to parry the complaints that I expect to be most widely shared, and remind the reader of the combination of virtues HOC displays, a combination not found in any rival view that I know of.

For many philosophers, my metaphysical starting point is baffling. If one is *not* inclined to endorse an empiricist/Humean approach to understanding things like causation or laws of nature, why do so when it comes to objective chance? Well, the restricted-Humean approach is what I try to motivate in the biggest chapter of the book, chapter 1. For many philosophers who think about chance, I am simply confused or mistaken about what the word 'chance' is supposed to mean: it is meant to refer to some sort of irreducible, single-case force (or tendency or propensity) that governs the unfolding of chancy events in our world—if there are any, that is.[1] For these philosophers, it is analytic that if our world is governed by deterministic laws, then there are no objective chances in it. What my analysis comes up with may, perhaps, deserve to be called a species of "objective probability"—a rival

---

[1] I don't think we really understand what could be meant by postulating primitive or propensity-type chances. At least, I am sure that I do not, and a large part of chapter 1 tries to persuade the reader that she does not, either.

to standard versions of frequentism, perhaps—but not an account of *objective chance.*

The purely terminological dispute is in the end not that important, although I stand by my view that common usage favors considering the chances of card games and roulette wheels as well as radioactive decays to be genuine, objective chances, with no implicit *caveat* concerning determinism of physical laws. Readers who view things differently are free to take HOC as a proposed account of objective probabilities which are (mostly, or perhaps always) not *chances.* What matters is that they be agreed to serve the same purpose that chances are famously supposed to serve—that of being, ideally, a "guide to life" in just the way codified in the Principal Principle (PP)—and to do so demonstrably.

The demonstration that HOCs can indeed play this role, in chapter 4, seems to me to be unassailable and to be a unique advantage that HOC has over all rival accounts of objective chance. Rival accounts permit at best question-begging or partial justifications of the PP, whereas HOC's unique features permit a justification that is *almost* air-tight, and one whose gaps or imperfections can be seen to be ones of no practical importance whatsoever. Until a rival account offers an even better justification of the PP, HOC stands alone in satisfying the most central desideratum of an account of chance.

At least, it does so if it does actually offer an account of what the objective chances are in the first place! Here is the second place where I expect many philosophers will have serious reservations about HOC: it may be thought to be open to the charge of being too vague to actually do the job of satisfactorily picking out the chance-facts in a complex world such as ours. Lewis' Best System Analysis of laws and chances is frequently criticized for the vagueness of the notions of simplicity, strength, and the idea of an optimal trade-off between them. HOC shares these sources of vagueness and apparently adds further sources, because of the explicitly pragmatic, user-friendly orientation discussed in chapter 3.

But this seeming vice turns out, on reflection, to be both inevitable and essentially harmless. The vagueness is mostly not about what, numerically speaking, the chance of A in setup S is; as I argued in chapter 3, the ordinary statistical and theoretical methods of inference used in science are clearly apt for determining HOCs, as long as we set aside inductive skepticism. And in this regard HOC is superior to rival accounts, such as propensity theories, that fail to give us a justification of the PP, because standard statistical methods clearly, if tacitly, presuppose the validity of the PP. Instead, the vagueness is about the borderline between the potential chance rules make it into the system and those that get excluded (for being redundant, or for enhancing strength too little in exchange for the loss of simplicity, or for sacrificing fit too much, and so on.) But this vagueness is sure to be harmless, since HOC is designed so as to ensure that both arguments of chapter 4 justifying the rationality of PP, the *a priori* and the consequentialist arguments, go through. A chance rule simply cannot (a) cover a large number of events in our world, and yet also (b) fail to serve as a good guide to life, either by failing to match actual frequencies or by misleading us about their stability and reliability. Thus, anything that we might judge to indeed deserve to be in the Best System is not going to lead us astray, unless induction itself fails us. Conversely, if we decide that a potential chance rule does not deserve to be included in the Best System, but *not* because of a problem of poor fit to the frequencies, we will know that it is still a statistical probability that we can use, for most purposes, just as if it were a full-fledged objective chance. And this, I believe, is a faithful reflection both of the way that probabilities are used in many areas of science (and other human activities), and of the vagueness of the boundaries of what we normally consider to be objective chances. In homage to the philosopher who set me on the path of trying to analyze objective chance, let me appropriate David Lewis' oft-stated rule about good conceptual analysis: vagueness in an analysis is no vice, when it matches the vagueness of the concept being analyzed.

Bibliography

Albert, D. Z. (1992). *Quantum Mechanics and Experience* (Vol. 46). Harvard University Press.

Albert, D. Z. (2000). *Time and Chance* (Vol. 114). Harvard University Press.

Albert, D. Z. (2015). *After Physics*. Harvard University Press.

Anscombe, G. E. M. (1971). *Causality and Determination*. Cambridge University Press.

Belot, G. (2016). Undermined. *Australasian Journal of Philosophy, 94*(4), 781–791. https://doi.org/10.1080/00048402.2016.1139604

Braddon-Mitchell, D. (2004). How do we know it is now now? *Analysis, 64*(3).

Callender, C. (2017). *What Makes Time Special?* Oxford University Press.

Cartwright, N. (1979). Causal laws and effective strategies. *Noûs, 13*(4), 419–437. https://doi.org/10.2307/2215337

Cartwright, N. (1983). *How the Laws of Physics Lie*. Oxford University Press. https://doi.org/10.1093/0198247044.001.0001

Cartwright, N. (1989). *Nature's Capacities and Their Measurement*. Oxford University Press.

Cartwright, N. (1999). *The Dappled World: A Study of the Boundaries of Science*. Cambridge University Press.

Clark, P. (2001). Statistical mechanics and the propensity interpretation of probability. In J. Bricmont & Others (Eds.), *Chance in Physics: Foundations and Perspectives*. Springer.

Correia, F., & Rosenkranz, S. (2018). *Nothing to Come: A Defence of the Growing Block Theory of Time*. Springer International. Retrieved from https://books.google.es/books?id=HLNZDwAAQBAJ

Cushing, J. T. (1994). *Quantum Mechanics: Historical Contingency and the Copenhagen Hegemony* (Vol. 27). University of Chicago Press.

de Finetti, B. (1937). La prévision: Ses lois logiques, ses sources subjectives. *Annales de l'Institut Henri Poincaré, 17*, 1–68.

de Finetti, B. (1990). *Theory of Probability* (Vol. 1). Wiley.

Diaconis, P. (1998). A place for philosophy? The rise of modeling in statistical science. *Quarterly of Applied Mathematics, 56,* 797–805. https://doi.org/ https://doi.org/10.1090/qam/99606

Dizadji-Bahmani, F., Frigg, R., & Hartmann, S. (2010). Who's afraid of Nagelian reduction? *Erkenntnis, 73*(3), 393–412.

Dupré, J. (1993). *The Disorder of Things: Metaphysical Foundations of the Disunity of Science.* Harvard University Press. Retrieved from https:// books.google.es/books?id=Ev3HvgSjb1EC

Eagle, A. (2010). *Philosophy of Probability: Contemporary Readings.* Routledge.

Earman, J. (1986). *A Primer on Determinism.* D. Reidel.

Earman, J., & Roberts, J. (1999). "Ceteris paribus," there is no problem of provisos. *Synthese, 118*(3), 439–478.

Elga, A. (2004). Infinitesimal chances and the laws of nature. *Australasian Journal of Philosophy, 82*(1), 67–76.

Emery, N. (2015). Chance, possibility, and explanation. *British Journal for the Philosophy of Science, 66*(1), 95–120.

Emery, N. (2017). A naturalist's guide to objective chance. *Philosophy of Science, 84*(3), 480–499. https://doi.org/10.1086/692144

Eriksson, L., & Hájek, A. (2007). What are degrees of belief? *Studia Logica, 86*(2), 185–215.

Fetzer, J. H., & Nute, D. E. (1979). Syntax, semantics, and ontology: A probabilistic causal calculus. *Synthese, 40*(3), 453–495.

Filomeno, A. (forthcoming). *Stable regularities without governing laws.* Studies in History and Philosophy of Modern Physics.

Ford, S. R. (2010). *What Fundamental Properties Suffice to Account for the Manifest World? Powerful Structure.*

Forrest, P. (2004). The real but dead past: A reply to braddon-Mitchell. *Analysis, 64*(4), 358–362.

Frigg, R. (2008). A field guide to recent work on the foundations of statistical mechanics. In *The Ashgate Companion to Contemporary Philosophy of Physics* (pp. 105–202). Routledge.

Frigg, R., & Hoefer, C. (2007). Probability in GRW theory. *Studies in History and Philosophy of Science Part B—Studies in History and Philosophy of Modern Physics, 38*(2). https://doi.org/10.1016/j.shpsb.2006.12.002

Frigg, R., & Hoefer, C. (2015). The Best Humean System for statistical mechanics. *An International Journal of Scientific Philosophy, 80*(Supplement 3), 551–574. https://doi.org/10.1007/s10670-013-9541-5

Frigg, R., & Werndl, C. (2011). Explaining thermodynamic-like behavior in terms of epsilon-ergodicity. *Philosophy of Science*, *78*(4), 628–652.

Giere, R. N. (1973). Objective single-case probabilities and the foundations of statistics. In P. Suppes & Others (Eds.), *Proceedings of the Fourth International Congress for Logic, Methodology and Philosophy of Science, Bucharest, 1971* (Vol. 74, pp. 467–483). Elsevier. https://doi.org/10.1016/S0049-237X(09)70380-5

Giere, R. N. (1976). A Laplacean formal semantics for single-case propensities. *Journal of Philosophical Logic*, *5*(3), 321–353.

Gillies, D. (2000). *Philosophical Theories of Probability*. Routledge.

Glynn, L. (2010). Deterministic chance. *British Journal for the Philosophy of Science*, *61*(1), 51–80.

Hájek, A. (1996). "Mises redux"—redux: Fifteen arguments against finite frequentism. *Erkenntnis*, *45*(2–3), 209–227.

Hájek, A. (2003). What conditional probability could not be. *Synthese*, *137*(3), 273–323.

Hájek, A. (2008). Interpretations of probability. In E. N. Zalta (Ed.), *Stanford Encyclopedia of Philosophy*. Urlhttp://Plato.Stanford.Edu/Archives/Spr2010/Entries/Probability-Interpret/.

Hájek, A. (2009). Fifteen arguments against hypothetical frequentism. *Erkenntnis*, *70*(2), 211–235.

Hall, N. (1994). Correcting the guide to objective chance. *Mind*, *103*(412), 505–517. https://doi.org/10.1093/mind/103.412.505

Hall, N. (2004). Two mistakes about credence and chance. *Australasian Journal of Philosophy*, *82*(1), 93–111.

Hawley, P. (2013). Inertia, optimism and Beauty. *Noûs*, *47*, 85–103.

Hoefer, C. (1997). On Lewis's objective chance: "Humean supervenience debugged." *Mind*, *106*(422), 321–334.

Hoefer, C. (2004). Causality and determinism: Tension, or outright conflict? *Revista de Filosofía (Madrid)*, *29*(2), 99–115.

Hoefer, C. (2005). Humean effective strategies. In P. Hájek, L. Valdés-Villanueva, & D. Westerståhl (Eds.), *Logic, Methodology and Philosophy of Science: Proceedings of the Twelfth International Congress*. KCL Press, 271–294.

Hoefer, C. (2011). Physics and the Humean approach to probability. *Probabilities in Physics*. https://doi.org/10.1093/acprof:oso/9780199577439.003.0012

Hoefer, C. (2018). "Undermined." *Undermined*. http://philsci-archive.pitt.edu/14886/.

Howson, C., & Urbach, P. (1993). *Scientific Reasoning: The Bayesian Approach.* Open Court. Retrieved from https://books.google.es/books?id=GB6_QgAACAAJ

Hughes, R. I. G. (1989). *The Structure and Interpretation of Quantum Mechanics.* Harvard University Press.

Humphreys, P. (2004). Some considerations on conditional chances. *British Journal for the Philosophy of Science, 55*(4), 667–680.

Ismael, J. (2008). Raid! Dissolving the big, bad bug 1. *Noûs, 42*(2), 292–307. https://doi.org/10.1111/j.1468-0068.2008.00681.x

Lavis, D. A. (2005). Boltzmann and Gibbs: An attempted reconciliation. *Studies in History and Philosophy of Science Part B: Studies in History and Philosophy of Modern Physics, 36*(2), 245–273.

Leitgeb, H. (2017). *The Stability of Belief: How Rational Belief Coheres with Probability.* Oxford University Press.

Levi, I. (1983). Review of studies in inductive logic and probability (ed. R. C. Jeffrey). *Philosophical Review, 92*(1), 116–121.

Lewis, D. (1973). Causation. *Journal of Philosophy, 70*(17), 556–567.

Lewis, D. (1980). A subjectivist's guide to objective chance. In R. C. Jeffrey (Ed.), *Studies in Inductive Logic and Probability* (Vol. II, pp. 263–293). University of California Press.

Lewis, D. (1986a). A subjectivist's guide to objective chance. In *Philosophical Papers* (Vol. II, p. 83). https://doi.org/10.1093/0195036468.001.0001

Lewis, D. K. (1986b). Chancy causation. *Philosophical Papers, 2*, 175–184.

Lewis, D. (1994). Humean supervenience debugged. *Mind, 103*(412), 473–490.

Lewis, D. (2000). Causation as influence. *Journal of Philosophy, 97*(4), 182–197.

Loewer, B. (2001). Determinism and chance. *Studies in History and Philosophy of Science Part B, 32*(4), 609–620.

Loewer, B. (2004). David Lewis's Humean theory of objective chance. *Philosophy of Science, 71*(5), 1115–1125. https://doi.org/10.1086/428015

Loewer, B., Weslake, B., & Winsberg, E. (n.d.). *Time's Arrows and the Probability Structure of the World.* Harvard University Press.

Lyon, A. (2010). Deterministic probability: Neither chance nor credence. *Synthese, 182*(3), 413–432.

Mackie, J. L. (1965). Causes and conditions. *American Philosophical Quarterly, 2*(4), 245–264.

Mackie, J. L. (1966). The direction of causation. *Philosophical Review, 75*(4), 441–466.

Maudlin, T. (2007a). *The Metaphysics within Physics* (Vol. 69). Oxford University Press.

Maudlin, T. (2007b). What could be objective about probabilities? *Studies in History and Philosophy of Science Part B, 38*(2), 275–291.

Maudlin, T. (2019). *Philosophy of Physics: Quantum Theory*. Princeton University Press.

McCall, S. (1994). *A Model of the Universe Space-Time, Probability, and Decision*. Oxford University Press.

Meacham, C. J. G. (2010). Two mistakes regarding the principal principle. *British Journal for the Philosophy of Science, 61*(2), 407–431.

Mellor, D. H. (1995). *The Facts of Causation* (Vol. 1). Routledge.

Miller, K. (2013). The growing block, presentism, and eternalism. In H. Dyke & A. Bardon (Eds.), *A Companion to the Philosophy of Time*, 345–364. Wiley-Blackwell.

Moore, N. (2017). The Sleeping Beauty Problem: What about Monday?. http://philsci-archive.pitt.edu/15029/.

Myrvold, W. C. (2012). Deterministic laws and epistemic chances. In Y. Ben-Menahem & M. Hemmo (Eds.), *Probability in Physics*, 73–85. Springer.

Myrvold, W. C. (2015). What is a wavefunction? *Synthese, 192*(10), 3247–3274.

Pollock, J. L. (1990). *Nomic Probability and the Foundations of Induction* (Vol. 102). Oxford University Press.

Pooley, O. (2013). Relativity, the open future, and the passage of time. *Proceedings of the Aristotelian Society, 113*(3pt3), 321–363.

Ramsey, F. P. (1926). Truth and probability. In A. Eagle (Ed.), *Philosophy of Probability: Contemporary Readings*, 48–71. Routledge.

Roberts, J. T. (2001). Undermining undermined: Why Humean supervenience never needed to be debugged (even if it's a necessary truth). *Proceedings of the Philosophy of Science Association, 2001*(3), S98–S108.

Russell, B. (1912). On the notion of cause. *Proceedings of the Aristotelian Society, 7*, 1–26.

Russo, F., & Williamson, J. (2011). Epistemic causality and evidence-based medicine. *History and Philosophy of the Life Sciences, 33*(4), 563–582.

Salmon, W. C. (1967). *The Foundations of Scientific Inference*. University of Pittsburgh Press. https://doi.org/10.2307/j.ctt5hjqm2

Schaffer, J. (2001). Causes as probability raisers of processes. *Journal of Philosophy, 98*(2), 75–92.

Sklar, L. (1993). *Physics and Chance: Philosophical Issues in the Foundations of Statistical Mechanics.* Cambridge University Press.

Sober, E. (2010). Evolutionary theory and the reality of macro probabilities. In E. Eells & J. H. Fetzer (Eds.), *The Place of Probability in Science* (Vol. 284), 133–160. Springer.

Strevens, M. (1999). Objective probability as a guide to the world. *Philosophical Studies, 95*(3), 243–275.

Strevens, M. (2011). Probability out of determinism. In C. Beisbart & S. Hartmann (Eds.), *Probabilities in Physics,* 339–364. Oxford University Press.

Suárez, M. (2011). Propensities and pragmatism. *Journal of Philosophy, 110*(2), 69–92.

Suárez, M. (2016). The Chances of Propensities. *British Journal for the Philosophy of Science,* 69(4), 1155–1177.

Suppes, P. (1970). *A Probabilistic Theory of Causality.* North-Holland.

Suppes, P. (1993). The transcendental character of determinism. *Midwest Studies in Philosophy, 18*(1), 242–257.

Uffink, J. (2006). *Compendium of the Foundations of Classical Statistical Physics.* In J. Butterfield & J. Earman (Eds.), *Philosophy of Physics,* 923–1074. Elsevier Press.

Van Fraassen, B. C. (1980). *The Scientific Image.* Oxford University Press.

Van Fraassen, B. C. (1989). *Laws and Symmetry* (Vol. 102). Oxford University Press.

van Lith, J. (2001). Ergodic theory, interpretations of probability and the foundations of statistical mechanics. *Studies in History and Philosophy of Modern Physics, 32*(4), 581–594.

von Mises, R. (1981). *Probability, statistics, and truth.* New York: Dover Publications. Retrieved from http://cataleg.ub.edu/record= b1208402~S1*cat

von Plato, J. (1981). Reductive relations in interpretations of probability. *Synthese, 48*(1), 61–75.

von Plato, J. (1982). The significance of the ergodic decomposition of stationary measures for the interpretation of probability. *Synthese, 53*(3), 419–432.

von Plato, J. (1983). The method of arbitrary functions. *British Journal for the Philosophy of Science, 34*(1), 37–47.

Wallace, D., & Timpson, C. G. (2010). Quantum mechanics on spacetime I: Spacetime state realism. *British Journal for the Philosophy of Science*, *61*(4), 697–727.

Werndl, C. (2011). On the observational equivalence of continuous-time deterministic and indeterministic descriptions. *European Journal for Philosophy of Science*, *1*(2).

Werndl, C., & Frigg, R. (2015). Rethinking Boltzmannian equilibrium. *Philosophy of Science*, *82*(5), 1224–1235.

Index

Note: figures and are indicated by *t* and *f* following the page number.

For the benefit of digital users, indexed terms that span two pages (e.g., 52–53) may, on occasion, appear on only one of those pages.